POLITICS AND PRESERVATION

STUDIES IN HISTORY,
PLANNING AND THE ENVIRONMENT

Series editors **The late Professor Gordon E. Cherry**
Professor Anthony Sutcliffe, *University of Leicester*

1 The Rise of Modern Urban Planning, 1800–1914
Edited by Anthony Sutcliffe

2 Shaping an Urban World
Planning in the twentieth century
Edited by Gordon E. Cherry

3 Planning for Conservation
An international perspective
Edited by Roger Kain

4 Metropolis 1890–1940
Edited by Anthony Sutcliffe

5 Arcadia for All
The legacy of a makeshift landscape
Dennis Hardy and Colin Ward

6 Planning and Urban Growth in Southern Europe
Edited by Martin Wynn

7 Thomas Adams and the Modern Planning Movement
Britain, Canada and the United States, 1900–1940
Michael Simpson

8 Holford
A study in architecture, planning and civic design
Gordon E. Cherry and Leith Penny

9 Goodnight Campers!
The history of the British holiday camp
Colin Ward and Dennis Hardy

10 Model Housing
From the Great Exhibition to the Festival of Britain
S. Martin Gaskell

11 Two Centuries of American Planning
Edited by Daniel Schaffer

12 Planning and Urban Growth in Nordic Countries
Edited by Thomas Hall

13 From Garden Cities to New Towns
Campaigning for town and country planning, 1899–1946
Dennis Hardy

14 From New Towns to Green Politics
Campaigning for town and country planning, 1946–1990
Dennis Hardy

15 The Garden City
Past, present and future
Edited by Stephen V. Ward

16 The Place of Home
English domestic environments
Alison Ravetz with Richard Turkington

17 Prefabs
A History of the UK Temporary Housing Programme
Brenda Vale

18 Planning the Great Metropolis
The 1992 Regional Plan of New York and its environs
David A. Johnson

19 Rural Change and Planning
England and Wales in the twentieth century
Gordon E. Cherry and Alan Rogers

Forthcoming titles

Planning Europe's Capital Cities
Aspects of nineteenth-century urban development
Thomas Hall

Selling Places
The marketing and promotion of towns and cities 1850–2000
Stephen V. Ward

POLITICS AND PRESERVATION

A Policy History of the Built Heritage, 1882–1996

JOHN DELAFONS

LONDON AND NEW YORK

First edition 1997
By E & FN Spon.
Reprinted 2004
by Spon Press,
2 Park Square, Milton Park, Abingdon, Oxon, OX14 4RN

Transferred to Digital Printing 2004

This book was commissioned and edited by Alexandrine Press, Oxford

Typeset in Great Britain by Cambrian Typesetters, Frimley, Surrey

ISBN 0 419 22390 8 (HB) 0 419 22400 9 (PB)

A catalogue record for this book is available from the British Library

Library of Congress catalog card number 97-66868

CONTENTS

ILLUSTRATIONS xi

ACKNOWLEDGEMENTS xiii

1. INTRODUCTION 1
   The Heritage 1
   Planning and Conservation 2
   Archaeology 2
   Previous Studies 2
   The Scope of Conservation 4
   The Process of Renewal 4

Part 1: 1882–1940 7

2. ORIGINS 9
   Early Examples 9
   The Victorian Restorers 13
   Ruskin, Morris and the SPAB 16
   A Dissenting Voice 21

3. ANCIENT MONUMENTS 23
   A Premature Initiative 23
   The First Legislation 23
   A European Comparison 25
   More Ancient Monuments Acts 29
   Other Developments 32

4. PLANNING 36
   Housing, Town Planning Etc. Act 1909 36
   Housing and Town Planning Act 1919 37
   Housing Etc. Act 1923 37
   Town and Country Planning Act 1932 38

5. DEMOLITION AND INACTION 1900–1940 43
Mediaeval and Tudor Buildings 43
Georgian and Victorian Buildings 44
A Resounding Protest 48
The Georgian Group 49
Intermission 51

Part 2: 1940–1975 53

6. PRELUDE 55

7. THE NEW SYSTEM 56
Town and Country Planning Act 1944 56
Town and Country Planning Act 1947 59

8. IMPLEMENTING THE NEW ACTS 62
Post-War Reconstruction 62
Listing Begins 65
Advisory Committee on Listing 65
Instructions to Investigators 66
Interim Lists 69

9. CONSERVATION FOR SOME 71
The National Trust 71
The Gowers Committee 72
The Historic Buildings Council for England 73
A Lone Critic 75

10. A POLICY VACUUM 77
Casework before Policy 77
Progress with Listing 78
Building Preservation Orders 79
Examples 79
Victorian Architecture 82
Two Landmark Cases 82
'Outrage' 85
A New Impetus 86

11. A NEW POLICY APPROACH 87
Urban Renewal 87

12. MINISTERS MAKE POLICY 89
Duncan Sandys 89
Richard Crossman 90

13. THE POLICY MOMENTUM 95
Civic Amenities Act 1967 95
'Historic Towns: Preservation and Change' 97
Four Studies in Conservation 98
Circular 53/67 100
The Town and Country Planning Act 1968 101
The Preservation Policy Group 102

14. CONTINUITY OF POLICY 104
The Department of the Environment 104
New Acts and New Circulars 104
Covent Garden 108

15. EUROPEAN ARCHITECTURAL HERITAGE YEAR 110
The Origins of EAHY 110
Objectives 111
Organization 111
Grants 112
Other Activities 113
Architectural Heritage Fund 113
What is Our Heritage? 114
EAHY Concludes 114
SAVE 115

Part 3: Churches 117

16. CHURCHES 119
Ecclesiastical Exemption 119
Church and State 120
Alterations 120
Redundancy 121
Archbishops' Commission on Redundant Churches 121
Faculty Jurisdiction Commission 124
Environment Committee 1986 125
Lord Skelmersdale's Statement 126
The Wilding Report 127
Cathedrals 128
Future Prospects 128

Part 4: 1976–1995 131

17. PRELUDE 133

18. ENGLISH HERITAGE IS BORN 136
A Consultation Paper 136
The Way Forward 137
National Heritage Act 1983 138

19. ENGLISH HERITAGE AT WORK 142
First Annual Report 142
Increasing Workload 143
An Advocacy Role 144
Public Image 146
Widening Scope 146

20. SELECT COMMITTEE 149
The Committee's Enquiry 149
Devil's Advocate 149
The Amenity Bodies 150
Victorian Buildings 151
Other Evidence 152
The Minister's Evidence 153
The Select Committee's Report 153
The Government Response 154

21. PRIORITIZING THE HERITAGE 156
The Department of National Heritage 156
English Heritage Reviewed 157
National Audit Office Report 157
A New Strategy 159
The Response 160
National Heritage Committee 161
The Government's Response 163

22. ADVICE AND DOCTRINE 165
Twenty Years of Circulars: 53/67 to 8/87 165
Criteria for Listing 167
Planning Policy Guidance 15 168
No 1 Poultry 172

23. SUSTAINABLE CONSERVATION 177
Sustainability 177
Attitudes to Conservation 178
The Heritage Industry 181
Maintenance and Repair 182
Policy Review 182
The Future 187

24. POSTSCRIPT: A CONSULTATION DOCUMENT 189

APPENDIX A: CHRONOLOGY 192

APPENDIX B: INSTRUCTIONS TO INVESTIGATORS 194

APPENDIX C: LISTING CRITERIA 201

APPENDIX D: STATISTICS 206

SELECT BIBLIOGRAPHY 208

INDEX 210

ILLUSTRATIONS

Cartoon by Alan de la Nougerade from *The Oldie*. (*Frontispiece*)

Woodstock Manor, Oxon in 1714. A royal palace for four centuries, it was demolished in 1723 despite Sir John Vanbrugh's attempt to save it. (*page 10*) (*By courtesy of the National Monuments Record*)

Sir John Vanbrugh (1664–1726). Architect and playwright. (Mezzotint by John Simon, after Sir George Kneller) (*page 11*) (*By courtesy of the National Portrait Gallery*)

Durham Cathedral. The East end rebuilt in 1798. One of what Pevsner calls James Wyatt's 'devastating improvements'. (*page 13*) (*By courtesy of the National Monuments Record*)

The Round Church, Cambridge (Holy Sepulchre): Ackerman's print of 1814 and after Salvin's alterations (1841). (*page 15*) (*By courtesy of the National Monuments Record*)

Highclere, Hants: after Charles Barry's rebuilding (1838), which enclosed the existing eighteenth-century house. (*page 16*) (*By courtesy of Country Life*)

Hadlow Court Castle, Kent: designed by the owner Walter Barton May (*c*. 1835) except for the tower designed by George Ladwell Taylor (1838). All but the tower was demolished in 1952. (*page 17*) (*By courtesy of the National Monuments Record*)

John Ruskin (left) with Dante Gabriel Rossetti, founder members of the SPAB. (Photo W.D. Downey (1863)) (*page 18*) (*By courtesy of the National Portrait Gallery*)

William Morris, founder and first secretary of SPAB. (Photo by Albert Lewis (1880)) (*page 19*) (*By courtesy of the National Portrait Gallery*)

Sir John Lubbock (Lord Avebury): progenitor of the first Ancient Monuments Act (1882). Contemporary cartoon by 'Figaro'. (*page 24*) (*By courtesy of the National Portrait Gallery*)

Stonehenge, Wiltshire, one of the sixty-eight monuments scheduled by the 1882 Act: in 1836, water colour by John Constable RA; in 1900 before Ministry of Works partial reconstruction; and as it is today. (*pages 26 and 27*) (*Water colour by courtesy of the Victoria & Albert Museum; photographs by courtesy of the National Monuments Record*)

The Clergy House (14th century), Alfriston, East Sussex. The first building acquired and restored by the National Trust (1896): before and after. (*page 33*) (*By courtesy of the National Monuments Record*)

The Piccadilly Hotel, Regent Street, London, by Norman Shaw (1905) begins the redevelopment of Nash's Regent Street. (*page 44*) (*By courtesy of the National Monuments Record*)

The Adelphi, London, by Robert Adam (1768). South front before (1932) and during demolition (1936), replaced by offices. (*page 45*) (*By courtesy of the National Monuments Record*)

Norfolk House, St James Square, London (1756): the ballroom. Demolished 1956 and replaced by offices. (*page 46*) (*By courtesy of the National Monuments Record*)

Carlton House Terrace, The Mall, London, by John Nash (1827). The Crown Estate Commissioners proposed to replace it by a department store in the 1930s but it survived this and a scheme to add several storeys of offices in 1946. (*page 48*) (*By courtesy of the National Monuments Record*)

Devonshire House, Piccadilly, London by William Kent (1733). The ballroom. Demolished 1924 and replaced by offices. (*pages 50 and 51*) (*By courtesy of the National Monuments Record*)

Holland House, Kensington (1605) after bomb damage in 1943. The remains were incorporated in a youth hostel. (*page 57*) (*By courtesy of the National Monuments Record*)

Exeter Cathedral: the Choir after Baedeker raid (1942). (*page 63*) (*By courtesy of the National Monuments Record*)

Stratford Place Nos 4–7, London, by R. Edwin (1773): subject of a building preservation order 1959. (*page 81*) (*By courtesy of London Metropolitan Archives*)

The Coal Exchange, City of London, by J.B. Bunning (1849); demolished for road widening 1962. (*page 83*) (*By courtesy of Country Life*)

The Euston Arch, London, by Philip Hardwick (1858): demolished 1962. (*page 85*) (*By courtesy of the National Monuments Record*)

Eldon Square, Newcastle on Tyne: before redevelopment (1966). (*page 92*) (*By courtesy of the National Monuments Record*)

The Shambles, Manchester: preserved in the midst of redevelopment. (*page 106*) (*By courtesy of the Press Association*)

Westminster Abbey before alterations to the West front: print by Hollar (1652). (*page 122*) (*By courtesy of the National Monuments Record*)

Westminster Abbey with Hawksmoor's west towers (1735): print from Maitland's *History of London* (1739). (*page 123*) (*By courtesy of the National Monuments Record*)

Telephone kiosk, Ladbrook Grove, London, by Giles Gilbert Scott: one of 1200 listed. (*page 147*) (*By courtesy of the National Monuments Record*)

Tudor tavern found inside Docklands warehouse, now relocated and restored as The Dickens Inn (*c.* 1980). (Photo Paul Berkshire) (*page 154*) (*By courtesy of the National Monuments Record*)

Albert Dock, Liverpool: before conversion to Tate Gallery of the North (1984). (*page 155*) (*By courtesy of the National Monuments Record*)

No 1 Poultry, City of London (1871): demolished for redevelopment as offices (1995). (*page 173*) (*By courtesy of the National Monuments Record*)

Windsor Castle, St George's Hall by Wyatville (1829): before and after the fire of 1993. (*pages 180 and 181*) (*By courtesy of The Royal Collection© Her Majesty the Queen*)

Uppark, West Sussex, by William Talman (1685–95): before the fire of 1989: restored by the National Trust. (*page 183*) (*By courtesy of the National Monuments Record*)

De La Warr Pavilion, Bexhill, East Sussex, by Erich Mendelssohn and Serge Chermayeff (1933–36): one of the first inter-war buildings to be listed. (*page 185*) (*By courtesy of The Architect*)

ACKNOWLEDGEMENTS

I acknowledge gratefully the help and encouragement that I have had from friends and colleagues, in particular Professor Barry Cullingworth, Hugh Corner, Hugh and Elizabeth Ellis-Rees, John Featherstone, Tony Fyson, Stuart Gilbert, Professor Malcolm Grant, Professor Peter Hall, Sir Desmond Heap, Mehdi Hodjar, Dr Crawford Knox, Professor Nathaniel Lichfield, Roger Suddards, Neil Summerton, Professor Anthony Sutcliffe, Professor John Worthington – none of whom is responsible for my errors or erroneous opinions. Lord and Lady Kennet were generous with their advice on churches and ecclesiastical exemption. Special thanks are due to Anne Woodward and Anna Eavis of the National Monuments Record, who gave invaluable help with my picture search, and to Yvonne Penford and Abi Gillet for their immaculate word-processing skills, Finally, the book would never have reached publication but for the shrewd advice and professional skill of Ann Rudkin.

I am grateful to the Controller of Her Majesty's Stationery Office for permission to reproduce extracts from the *Instructions to Investigators* of 1944, printed in Appendix B; and for permission to print extracts from Departmental Circulars in Appendix C; and to Penguin Books Ltd for permission to quote from Richard Crossman's *Diaries of a Cabinet Minister* in chapter 12. And I am grateful to The Royal Collection© Her Majesty the Queen, *Country Life*, the London Metropolitan Archives, the National Monuments Record, the National Portrait Gallery, *The Oldie*, and the Victoria and Albert Museum for permission to use illustrations.

An earlier version of chapter 4 appeared in the June 1994 issue of the *Journal of Planning and Environmental Law*.

John Delafons
1996

For Sheila

'Bugger me – this is a conservation zone!'

1

INTRODUCTION

THE HERITAGE

The word 'heritage' has become a very capacious portmanteau. *The Ancient Monuments Protection Act 1882* listed twenty-nine monuments in England and Wales (plus twenty-one in Scotland and eighteen in Ireland).[1] It was the first such statutory list. Today there are some 15,000 scheduled monuments in England, over 500,000 listed buildings, and over 9,000 conservation areas. English Heritage estimates that there are about 600,000 archaeological sites, and a new inventory is being compiled. The statutory lists include, along with buildings of architectural distinction or historic significance, such items as dog kennels, lamp posts, bollards, stiles, pillar boxes, telephone kiosks, railings, fences and grave watchers' huts. English Heritage is also compiling lists of listable post-World War II buildings, historic gardens and battlegrounds. In recent years the popular concept of 'the heritage' has been extended to cover an enormous variety of buildings and artefacts that bear little or no relationship to the traditional statutory term 'buildings of special architectural or historic interest'.

How did we arrive at this position? At first conservation was not an issue adopted with enthusiasm by government and they were reluctant to get involved. But in time the heritage was found to be a source of political capital. The values that motivated the pioneers became obscured by other interests. What began as an antiquarian and scholarly pursuit, and became an elitist cause, has developed into a populist movement. As attitudes changed, so did government policy. Such developments in policy usually reflect wider social or cultural influences. They do not originate solely within government. Now some questions are being asked about the scope and purpose of conservation and its demand on resources. The conflict between the desire to preserve and the need to change generates political tensions.

This study traces the policy history of urban conservation and sets it in its political context. The main theme is to show how governments have responded to the growing interest in the heritage. It focuses on the legislative and administrative aspects of the subject. This may seem a narrow vein to work in the deep mine that is conservation. But conservation cannot be pursued solely by private initiative. It requires intervention by the State. Those concerned for the heritage and students of conservation need to know how the present system evolved and how the legislative process works. My hope is that those involved with, or concerned for, conservation will find this account both interesting and useful.

Planning and Conservation

Conservation predates town and country planning in its legislative form, but became assimilated into the planning system and in some respects came to dominate it. In the first twenty years or so after the Town and Country Planning Act 1947 conservation was an integral but relatively minor part of the planning system.[2] In the subsequent twenty years it became increasingly prominent and acquired a life of its own. Indeed 'planning' and 'conservation' are sometimes portrayed as antagonists. It has become important to reassert that conservation is part of planning and that planning must incorporate conservation objectives. But, just as planning is wider in scope than conservation, so conservation goes wider than planning. It is the relationship between the two that needs unravelling, and the historical approach can help to do this.

There can be no mistaking the triumphalist tone of conservationists today. Michael Ross, the former head of Listing Branch in the Department of the Environment, proclaims in the opening chapter of his book *Planning and the Heritage*, 'Politicians have ignored conservation at their peril . . . the idea of conservation, the presumption that the old must survive – and on occasion adapt – has triumphed . . . The philosophy of conservation . . . *is* accepted by the majority; architects, developers, planners and politicians cannot ignore this fact.'[3] If one has reservations about the conservationist cause it is not because one ignores 'this fact' but because it is a more complex issue than is often assumed. This too will become apparent as the story unfolds.

Archaeology

This study of urban conservation does not deal fully with the parallel history of British archaeology. It is a curious fact that, although the legislative history begins with ancient monuments, in subsequent years ancient monuments and town and country planning follow separate paths, and the relationship between them has never been fully resolved, although Planning Policy Guidance Note No. 16 seeks to clarify it.[4] Archaeology has developed a much more scholarly bias than planning, which in its origins was concerned chiefly with new development and, as this study shows, has only slowly been adapted to the needs of conservation. Archaeology developed its own culture and methods quite distinct from planning, and this has been reflected in legislation. For many years some departmental rivalry existed between those concerned with ancient monuments and those responsible for listed buildings. While those functions were brought together in 1970 in the new Department of the Environment, they were never fully integrated; and since the formation of the Department of National Heritage in 1992 new problems of coordination have been generated. This element of Departmental rivalry adds a certain spice to the story.

Previous Studies

I have not found that any previous writer on conservation has approached the subject in the way that I propose. Most concentrate either on promoting the conservationist cause or on the technical and legal aspects of conservation. Some authors include a brief account of its origins and an outline of previous legislation, notably Wayland Kennet in *Preservation*[5] and Alan Dobby in *Conservation and Planning*[6] – two of the best books on the subject, that take the story up to the 1970s. But none deals with the legislative and policy history in detail, or takes as the principal theme the government response to changing attitudes towards historic buildings and their conservation. One of the most interesting books on the subject is *The Future of the Past* edited by Jane Fawcett.[7] This was published in 1976 under the auspices of the Victorian Society and derived in part from an exhibition sponsored by the Society at the Central School of Art and Design in London in 1971 (and subsequently acquired by the Victoria and Albert Museum). But, despite its subtitle 'Attitudes to Conservation 1174–1974', it does not attempt to provide a continuous or comprehensive account of the history of conservation but comprises a collection of brilliant essays by authors as diverse as Nikolaus Pevsner and Mark Girouard on the one hand, and John Betjeman and Osbert Lancaster on the other. The introductory chapter by Nikolaus Boulting, however, provides a short summary of the progress of conservation in Britain from the earliest days to the Act of 1968 and I have drawn on this for two or three early examples.

The weightiest study of architectural conservation that I have come across is the unpublished PhD thesis by Jukha Ilmari Jokilehto, completed in 1986 for the Institute of Advanced Architectural Studies at the University of York. Its full title indicates its scope and purpose: *The Contribution of English, French, German and Italian Thought towards an International Approach to the Conservation of Cultural Property*. Although it extends much wider than my own study, it includes much fascinating material on the course of conservation in England during the eighteenth and nineteenth centuries. It is a most remarkable piece of work but it does not deal with the policy and legislative history of conservation which is the subject of the present study. Mr Jokilehto later became European Director of the International Council for the Conservation of Monuments and Sites.

There is an enormously rich literature, of course, on architectural history and, in the past thirty years or so, a rapidly diversifying stream of studies dealing with conservation from the aesthetic, psychological, sociological and anthropological perspectives.[8] But my main sources have been the Official reports of Debates in Parliament (Hansard), Parliamentary papers, official publications and Departmental Bill Papers at the Public Records Office (the last far less informative than might be thought). There is an excellent bibliography which takes the subject up to 1976 – *A Critical Bibliography of Building Conservation* by J.F. Smith,[9] but this does not deal with the legislative history or policy development. It is interesting to note that Smith, having surveyed the literature in great detail, remarks that conservation is 'now generally recognised as an element in every aspect of architectural and urban design', whereas 'two decades ago conservation was regarded as a minority interest.' I have not attempted to extend Smith's bibliography but I have recorded in the notes to each chapter and in the Select Bibliography the books that I have found most useful or interesting, and these will serve as suggestions for further reading.

It has not been possible to take account of *Preserving the Past: The Rise of Heritage in Modern Britain* edited by Michael Hunter[10] which appeared too late for this study. The book comprises essays by nine contributors on aspects of conservation history and includes new material, notably on the early days of the Society for the Preservation of Ancient Buildings, the evolution of the ancient monuments regime and the early days of listing, together with an up-to-date bibliographical essay by the editor. It does not deal in detail with the political and policy aspects of the subject, which are the main concern of the present study.

The Scope of Conservation

The fact that conservation has a long legislative history does not mean that its objectives have remained unchanged. The scope of conservation has widened enormously over the past hundred years and diverse influences have shaped its development. The history of conservation over that period will reveal that those early conservationists who started it all have been joined by a great many other participants. What was once the concern of antiquarians and scholars is now only part, and perhaps only a rather esoteric part, of a much wider and more complex aspect of modern life that involves questions of national identity and personal psychology.[11] These are deep waters.

It is not the intention of this study to provide a detailed guide to the legal aspects of conservation. The late Roger Suddards's admirable textbook on the subject *Listed Buildings* lists over sixty relevant Acts, and there has been further legislation since his book first appeared in 1988 (published in its third edition in 1996).[12] I record the main legislative landmarks but my purpose is to examine how the policy behind the legislation has evolved and how the concept of safeguarding such buildings has changed from restoration to preservation, and from protection to conservation, and how it may now be changing again. The significance of these linguistic distinctions will become apparent but they have not been used consistently in the past and they overlap to a considerable extent. In brief, however, restoration implies significant work on the fabric, which may extend to substantial rebuilding; preservation implies retention with minimal alteration; protection implies safeguards against demolition or ill-advised improvement but does not exclude possible adaptations or alterations; finally conservation goes much wider than the protection of individual buildings and can extend to whole areas and to features other than buildings. It also implies a different policy approach, reflecting a broader range of public interest. Those distinctions are sufficient for our present purpose and already indicate that there is a diversity of motives or objectives involved. Conservation, being broadest in scope, provides a convenient inclusive term for the whole business. I have used 'preservation' in my title solely in the interests of alliteration.

The Process of Renewal

The need to renew and rebuild has obviously been a fact of life since man first began to build. It is often problematical to distinguish historically between the need to rebuild for purely practical reasons and the desire to rebuild for the greater glory of God or for the greater glory of the proprietor. Clearly religious motives played a part for the former and secular motives for the latter. Perhaps the answer is that, when the practical need or opportunity to rebuild presented itself, then aesthetic considerations also took effect.

John Harvey in *Conservation of Buildings* gives a detailed account of how works of restoration were carried out to important buildings from mediaeval to Tudor times.[13] These were not motivated simply by the need for repairs but often reflected reverence and respect for buildings that had survived for centuries. It is hardly necessary to adduce evidence of the historic process of renewal, since it is apparent in almost all buildings that survive from earlier than the nineteenth century and not only among ecclesiastical buildings. At royal palaces, aristocratic mansions, country gentlemen's homes, merchants' town houses and countless other buildings, the work of rebuilding, additions and enhancement continued over the centuries. Considerations of practical necessity, aesthetic interests and personal vanity provided the driving force. In all of this there seems to have been a sense of progress and innovation, whether in the wealthy plutocrat abandoning the ancestral home of his Tudor forbears and building himself a Palladian mansion, or in the successful merchant cladding his half-timbered house with a new brick facade. At some point, however, the process of rebuilding turned into restoration, which led in turn to the demand for preservation.

It is difficult to identify that precise turning point. Sir Christopher Wren encountered much criticism when he produced his plans for restoring old St. Paul's, which included taking down the tower and adjoining bays in the nave, choir and transepts, and remodelling the whole of the central area to support a huge dome.[14] But the dispute was based on liturgical rather than antiquarian grounds. That dispute persisted when Wren proposed an entirely new design after the Great Fire but he eventually prevailed. He might not have done so today, if the response to the fire at Windsor Castle in 1992 reflects current attitudes to conservation (see chapter 23).

Outline

The approach that I have adopted is mainly and deliberately chronological. Part 1 deals with the origins of conservation and its cultural background, leading to the early legislation of 1882 and forward to the end of the 1930s. Part 2 deals with the post-war Planning Acts and the ever-broadening scope of conservation up to 1975, European Architectural Heritage Year, which for many at that time marked the high-tide of conservation in Britain. Part 3 deals separately with churches, which are subject to distinctive ecclesiastical procedures and are perhaps now the most endangered part of the heritage. Part 4 brings the story up to the present time and the attempt to apply Thatcherite precepts to conservation. The concluding chapter introduces a new concept: sustainable conservation. A Postscript deals with the government's new consultation document *Protecting Our Heritage* published in June 1996.

Note: This study deals primarily with England. Wales has the same legislation and similar policies. Scotland has separate legislation.

NOTES

1. 4524 Vict. c.73.
2. 10 and 11 Geo 6 c.51.
3. Ross, M. (1996) *Planning and the Heritage: Policy and Procedure*, 2nd ed. London: E. & F.N. Spon.
4. Department of the Environment (1990) *Planning Policy Guidance 16: Archaeology and Planning*. London: HMSO.
5. Kennet, W. (1972) *Preservation*. London: Temple Smith.
6. Dobby, A. (1978) *Conservation and Planning*. London: Hutchinson.
7. Fawcett, J. (ed.) (1976) *The Future of the Past. Attitudes to Conservation 1740–1974*. London: Thames & Hudson.
8. Lowenthal, D. (1985) *The Past is a Foreign Country*. Cambridge: Cambridge University Press. The more recent epistemological approach to the past is reflected in Samuel, R. (1994) *Theatres of Memory*. London: Verson.
9. Smith, J.F. (1978) *A Critical Bibliography of Building Conservation*. London: Mansell.
10. Hunter, M. (1996) *Preserving the Past: The Rise of Heritage in Modern Britain*. Stroud Alan Sutton Publishing.
11. Lowenthal, D. and Binney, M. (eds.) (1981) *Our Past Before Us: Why do We Save It?* London: Temple Smith.
12. Suddards, R.W. and Hargreaves, J.M. (1996) *Listed Buildings: The Law and Practice of Historic Buildings, Ancient Monuments an Conservation Areas*, 3rd ed. London: Sweet and Maxwell.
13. Harvey, J. (1972) *Conservation in Buildings*. London: John Baker, pp.157–175.
14. Summerson, J. (1963) *Heavenly Mansions and Other Essays*, new ed. New York: W.W. Norton.

PART 1
1882–1940

2

ORIGINS

EARLY EXAMPLES

Several writers on the subject of conservation have sought to trace its earliest manifestations in Britain but few have been able to follow it much earlier than the eighteenth century. Michael Ross identifies the 'first glimmerings' in the latter part of the seventeenth century and cites John Aubrey's prolific but chaotic antiquarian studies (none of which was published until after his death in 1697) and Anthony a Wood's *Historia et Antiquitates Univ. Oxon.* (1674) as displaying an early interest in ancient monuments and mediaeval English architecture.[1] But these are not relevant to the process of conservation as distinct from scholarly pursuits.

Nikolaus Boulting cites two notable examples of governmental intervention that are of singular interest.[2] In 1560 Queen Elizabeth issued a proclamation forbidding the 'defacing of monuments of antiquity, being set up in the churches or other public places for memory, and not for superstition'. This was prompted by reaction against the post-Reformation desecration of churches and other buildings, but the Queen was concerned for the sensibilities of her subjects rather than for the antiquarian interest of the memorials themselves. By contrast in 1641, under the Commonwealth, the House of Commons ordered that 'Commissions be sent into all countries, for the Defacing, Demolition, and quite taking away of all Images, Altars or Tables turned Altarwise, Crucifixes, superstitious Pictures, Monuments, and Relicts of idolatry, out of all churches and chapels'. In both cases, the Elizabethan and Cromwellian, the motive was political as much as religious. The significance of these two early examples is that they are probably the first instances (except in times of civil war and at the Dissolution) of conservation or demolition being the subject of deliberate government policy.

There were important topographical surveys in the sixteenth and seventeenth centuries, including John Stowe's Survey of London (1598), the works of Camden and Leland, and county histories such as Lambarde on Kent and Carew on Cornwall, which recorded noteworthy buildings and sometimes ruins.[3] John Harvey cites the documentary studies of architecture by Roger Dodsworth and Sir William Dugdale published in 1665–73 and *A Survey of the Cathedrals* by Brown Willis in 1727–30.[4]

This antiquarian interest led to the founding of the Society of Antiquaries in 1717 which, although chiefly interested in archaeology and ancient documents, also maintained an interest in mediaeval architecture.[5] The Society still exists and is the oldest of the bodies concerned with conservation. Cultivated taste in the eighteenth century, however, was more concerned with classical architecture and new building than with either archaeological remains or early English buildings. The growing interest in local history and historical remains did not generally reflect a concern for their preservation.

One early and often quoted example of concern for historic buildings is of exceptional interest both on account of its author and because of its recognition of the variety of factors that underlie the conservation instinct. Sir John Vanbrugh, writing to the Duchess of Marlborough on 11 June 1709, pleads for the preservation of Woodstock Manor:[6]

There is perhaps no one thing, which the most Polite part of Mankind have more universally agreed in; than the Vallue they have ever set upon the Remains of distant Times. Nor amongst the Severall kinds of those Antiquitys, are there any so much regarded, as those of Buildings; Some for their Magnificence, or curious workmanship; and other; as they move more lively and pleasing Reflections (than History without their Aid can do) on the Persons who have Inhabited them; On the Remarkable things which have been transacted in them, Or the extraordinary Occasions of Erecting them.

That short passage reflects everything implied by the statutory phrase 'buildings of special architectural or historic interest', but statutory provision for buildings such as Woodstock Manor had to wait more than two hundred years until the introduction of building preservation orders in the Ancient Monuments Act of 1913. Unfortunately Vanbrugh's intervention did not save Woodstock Manor, which fell into disuse and is now barely visible.

The fact that Vanbrugh was not alone at that time in expressing a feeling for ancient buildings is attested by the following extract from an article in one of the many early eighteenth-

Woodstock Manor, Oxon in 1714. A royal palace for four centuries, it was demolished in 1723 despite Sir John Vanbrugh's attempt to save it.

Sir John Vanbrugh (1664–1726). Architect and playwright. (Mezzotint by John Simon, after Sir George Kneller)

century periodicals *Common Sense* in 1739 to which John Cornforth has drawn attention:[7]

There was something respectable in those old hospitable Gothic Halls, hung around with helmets, breast-plates, and swords of our ancestors: I entered them with a constitutional sort of reverence, and looked upon those arms with gratitude as the terror of former ministers and the check of kings . . . and when I see them thrown by to make way for some tawdry gilding and carving, I can't help considering such alterations as ominous even to our constitution. Our old Gothic constitution had a noble strength and simplicity in it, which was well represented by the bold arches and the solid pillars of the edifices.

Such sentiments reflect a respect for antiquity and a distaste for modernity, a common conservative instinct, rather than an active concern for conservation. It is a persistent element in the English character, perhaps in the culture of all modern societies, which eventually leads to a demand for government action to preserve the relics of the past. But such action did not start for another hundred and fifty years.

Towards the end of the eighteenth and in the early nineteenth centuries there was a spate of elaborate and scholarly illustrated books that celebrated the architectural achievements of the Georgian and Regency periods. These included T. Malton *Picturesque Tour through the Cities of London and Westminster* (1790); A.C. Pugin and T. Rowlandson *The Microcosm of London*, 3 volumes (1808–9); J. Britton and A.C. Pugin *Illustrations of the Public Buildings of London*, 2 volumes (with plans, sections and elevations) (1825); D. Lyons *Environs of London*, 4 volumes (1792–96); J. Hassell *Views of Seats* (1804); and many more.[8] These were predecessors of the 'coffee table' books that have remained a staple of the publishing industry. While they were not explicitly concerned with conservation, they no doubt encouraged and disseminated a wider interest in English architecture and an awareness of its historical development. But despite this interest in architecture, there was still little demand for the conservation of buildings from earlier periods.

Archaeology attracted more general interest than historic buildings among the early Victorian audience. Since the Society of Antiquaries remained a select learned institution, the more popular interest was met by the foundation in 1842 of the British Archaeological Association, which was promptly torn apart by personal antipathies, resulting in the formation of the Archaeological Institute of Great Britain in 1844.

France seems to have had the honour of starting the listing business. In 1830 the French government set up the Commission des Monuments Historiques, which in 1837 began to compile an inventory of historic buildings. It was estimated that, as originally planned, this task would require nine hundred volumes and take two hundred years to complete. Then Prosper Merimée took over in 1840 and published the first list of fifty nine monuments, all mediaeval or earlier.[9] The work then continued at a moderate pace and is not yet complete.

England had to wait until 1882 for listing to begin. But there is some evidence of an interest in the preservation of historic buildings developing earlier in the nineteenth century. The essayist Leigh Hunt in *The Old Court Suburb* (1857) deplores the rumoured demolition of Holland House in Kensington, which he describes as 'the only important mansion, venerable for age and appearance, which is now to be found in the neighbourhood of London'.[10] There were said to be plans to turn it into a workhouse. Leigh Hunt exclaims 'every feeling of memory seems to start up at the threat and cry No! No!' In fact, not for the last time in the history of conservation, the 'threat' proved to be illusory and Hunt found that the noble owner was engaged in carrying out beneficial works of restoration.

There is probably other material lurking around in unread books that would provide material for an anthology of conservation sentiment from around mid-Victorian times, but one further example must suffice to show the trend of opinion. In 1860 T.C. Croker in *A Walk from London to Fulham* paid tribute to 'Mr. C.J. Richardson, an architect, whose publications illustrative of Tudor Architecture and domestic English antiquities have materially tended to diffuse a feeling of respect for the works of our ancestors, and to forward the growing desire to preserve or restore edifices which time and circumstances have spared this country.'[11] It does not seem that this 'feeling of respect for the works of our ancestors', and 'the growing desire to preserve or restore edifices which time and circumstances have spared this country', were very widespread at that time, but concern for ancient buildings took off and accelerated in the second half of the nineteenth century.

The catalyst, paradoxically, seems to have been the strenuous and pervasive efforts of those architects and their patrons who embarked on the 'restoration' of mediaeval churches. Their often drastic methods eventually provoked a radical reaction from those who wanted to preserve such buildings from the restorers.

The Victorian Restorers

Towards the end of the eighteenth century the movement began that led to the extraordinary passion for rebuilding English cathedrals, and lesser churches, that lasted throughout the nineteenth century. But from the first those activities attracted fierce criticism from the Society of Antiquaries and others. James Wyatt (d.1813) was the first to practice this type of ecclesiastical work on a large scale and was much in demand by Deans and Chapters. He undertook dramatic works of reconstruction to both exteriors and interiors of many major buildings including the cathedrals of Lichfield (the first, in 1788), Salisbury, Durham and Westminster Abbey. But his work did not go uncriticized. John Carter was

Durham Cathedral. The East end rebuilt in 1798. One of what Pevsner calls James Wyatt's 'devastating improvements'.

also an architect and adviser to the Society of Antiquaries. He used the columns of the *Gentleman's Magazine* to berate Wyatt's activities. Carter caught up with him at Durham in 1796 where Wyatt had already got as far as rebuilding the east end and had replaced the Perpendicular window of the east transept with an Early English rose window of his own design, added new turrets and the present North Porch.[12] But at that point Carter stopped him in his tracks before he could destroy Bishop Neville's reredos, add an octagonal tower or demolish the late Norman Galilee Chapel, as he had intended. Wyatt combined the appetite for demolition, that the Victorian restorers later displayed, with an eighteenth century taste for rebuilding in his own eclectic manner. Pugin called him 'this monster of architectural depravity'.[13] But no-one after Wyatt had the free hand that he enjoyed. His successors were the Camden Society, Ruskin and the Society for the Protection of Ancient Buildings (SPAB), as we will find in tracing the course of conservation.

The urge to 'restore' stemmed partly from a well-justified concern about the decrepit state of many mediaeval churches, partly from the Victorian sense of religious revival, and partly from perverse aesthetic theory. The villains were the Camden Society, founded in 1839 by two undergraduates of Trinity College, Cambridge, J.M. Neale and Benjamin Webb, with their tutor the Rev. T. Thorp as chairman. They stated the Society's object as 'to promote the study of Ecclesiastical Architecture and Antiquities, and the restoration of mutilated Architectural remains'. In the four short years of the Society's existence it rapidly came to exercise extraordinary influence, with its pamphlet *A Few Words to Church-builders* and its monthly journal *The Ecclesiologist*. Their original aim was to ensure that new churches, or churches undergoing refurbishment, were suitable for the liturgical practices that they favoured. But this soon led them to adopt a very doctrinaire approach to church restoration. They insisted on vetting the working drawings of schemes submitted for their approval. Their philosophy was stated in an early issue of *The Ecclesiologist* as follows: 'We must, whether from existing evidence or from supposition, recover the original scheme of the edifice as conceived by the first builder, or as begun by him and developed by his immediate successors.'[14] Observing that many churches and cathedrals had been altered in a variety of styles, they urged that they should be restored to their 'purest' form – which they took to be the 'middle pointed' or Decorated form of English Gothic. This approach often led to a process of radical rebuilding, sometimes from the ground up.

In 1845 the Cambridge University authorities, following a furious theological dispute, insisted that the Camden Society be dissolved. But it was at once reborn as the Ecclesiological Society and continued its work with undimmed ardour. Kenneth Clark in *The Gothic Revival* sums up their achievements: 'For fifty years almost every new Anglican church was built and furnished according to their instructions; that is to say, in a manner opposed to utility, economy and good sense – a very wonderful achievement in the mid-nineteenth century.'[15]

The Ecclesiological Society's influence was just as strongly felt in church restoration as in new building, and the campaign was carried forward and widened in style by leading architects such as Salvin, Bodley, Pearson and, most notably, Sir Giles Gilbert Scott. Scott's labours were prodigious. A list published in *The Builder* in 1878 gives the names of over 730 buildings with which he was concerned after 1847, including 39 cathedrals, 476 churches, 25 colleges or college chapels and a host of secular buildings: the list is said to be incomplete. Jane Fawcett, in her detailed assessment of the Victorian restorers' extraordinary assault on the English cathedrals, acknowledges the careful research on which much of Scott's work

The Round Church, Cambridge (Holy Sepulchre): Ackerman's print of 1814 and after Salvin's alterations (1841).

was based and his superb architectural skills. But she cannot forgive the over-confident manner in which he swept away much of the mediaeval fabric.[16]

While the church restorers, driven by liturgical as well as aesthetic motives, were busy about their business, the urge to alter and rebuild historic buildings was also infecting the owners of country houses. During the nineteenth century, stimulated by economic prosperity, new wealth and the need to cater for the elaborate country-house social life, many houses dating from Tudor to Georgian periods were given a radical face-lift or major extensions. While some of this work was clumsy and even grotesque, Mark Girouard observes that much of it was skilfully done by such architects as Salvin, William Burn and George Devey.[17] Scholarly restoration work of high quality was undertaken. Georgian Gothic was generally discarded, but nothing by Vanbrugh and only a few good Palladian buildings were lost. In some cases previous alterations were removed and earlier features restored, as where Georgian sash windows in Elizabethan houses were replaced with stone mullions and transoms. Eventually much of this Victorian work was demolished as owners struggled to adapt their property to more straitened circumstances. Thus the evidence for this phase of renewal was depleted, but there are sufficient examples remaining to moderate the sense of loss.

Highclere, Hants: after Charles Barry's rebuilding (1838), which enclosed the existing eighteenth-century house.

Ruskin, Morris and the SPAB

By far the most passionate voice to be raised against the ecclesiastical restoration movement was that of Ruskin. His early work *The Seven Lamps of Architecture* first appeared in 1849 when Ruskin was only thirty: he lived another fifty years.[18] The sixth of the seven lamps was the Lamp of Memory, and here Ruskin proclaims his anti-restoration creed in the kind of violent language which sometimes erupts like a shark's fin above the surging ocean of his prose style. He says that 'the preservation of the buildings we possess' does not belong to his present purpose but he feels that he must speak 'a few words' on the subject 'as especially necessary in modern times'. Then he launches into five pages of sustained attack on the practice of restoration:

> It means the most total destruction which a building can suffer: a destruction out of which no remnants can be gathered: a destruction accompanied with false description of the thing destroyed. Do not let us deceive ourselves in this important matter; it is *impossible*, as impossible as to raise the dead, to restore anything that has ever been great or beautiful in architecture . . . Do not let us talk then of restoration. The thing is a Lie from beginning to end.

Hadlow Court Castle, Kent: designed by the owner Walter Barton May (*c.* 1835) except for the tower designed by George Ladwell Taylor (1838). All but the tower was demolished in 1952.

Having comprehensively condemned restoration, Ruskin turns to the argument that restoration may be a necessity because of the dilapidated state of the building. He has his answer to that.

> Granted. Look the necessity full in the face, and understand it on its own terms. It is a necessity for destruction. Accept it as such, pull the building down, throw its stones into neglected corners, make ballast of them, or mortar, if you will; but do it honestly, and do not set up a Lie in their place.

He then develops his alternative approach to preservation (and he may have been the first to make this distinction between restoration and preservation): 'Take good care of your monuments, and you will not need to restore them.' But he urges only the minimum of repair and maintenance: 'do not care about the unsightliness of the aid: better a crutch than a lost limb'. He concludes by affirming that there can be no question of whether or not to preserve 'the buildings of past times': seizing italics he declares '*We have no right whatever to touch them.*'

Ruskin attempted to initiate some practical steps when in 1854 he asked the Society of Antiquaries to set up a Fund 'to be subscribed for the preservation of Medieval (*sic*) Buildings'. He offered to subscribe £25 himself but the Fund never attracted much support, although the Society occasionally intervened to prevent demolition, as with the Guildhall at Worcester in 1876.[19] In 1855 Ruskin persuaded the Society to circulate to all its Fellows a brief but cogent memorandum on restoration drawn up by its Executive Committee and signed by their President, Earl Stanhope. It is worth quoting in full since the principles that it enunciated were very similar to those which Morris produced in drawing up his Manifesto for the Society for the Protection of Ancient Buildings more than twenty years later and which has tended to overshadow the earlier initiative by the Society of Antiquaries:[19]

The numerous instances of the Destruction of the character of Ancient Monuments which are taking place under the pretence of Restoration, induce the Executive Committee, to which the Society of Antiquaries has entrusted the management of its 'Conservation Fund', to call the special attention of

John Ruskin (left) with Dante Gabriel Rossetti, founder members of the SPAB. (Photo W.D. Downey (1863))

the Society to the subject, in the hope that its influence may be exerted to stop, or at least moderate, the pernicious practice.

The evil is an increasing one; and it is to be feared that, unless a strong and immediate protest be made against it, the monumental remains of England will, before long, cease to exist as truthful records of the past . . .

The Committee strongly urge that, except where restoration is called for in Churches by the requirements of Divine Service, or in other cases of manifest public utility, no restoration should ever be attempted, otherwise than as the word 'restoration' may be understood in the sense of preservation from further injuries by time or negligence: they contend that anything beyond this is untrue in art, unjustifiable in taste, destructive in practice, and wholly opposed to the judgement of the best Archaeologists.

The strange thing is that, despite Ruskin's onslaught on the restoration industry and

William Morris, founder and first secretary of SPAB. (Photo by Albert Lewis (1880))

despite his growing reputation, the business of restoration continued apace for the next thirty years. Although Ruskin converted the Society of Antiquaries to his view, it was not until William Morris, an altogether more powerful and practical character, took up the cause that the extravagant vandalism of the restorers was arrested. Morris had himself been an exponent of restoration and his own firm had supplied quantities of furnishings and stained glass for the purpose. But while driving near Kelmscott on 4 May 1876 he was horrified to see Burford Church being demolished in readiness for restoration. In the following year he saw similar activity at the Minster in Tewkesbury, under Gilbert Scott's direction, and this prompted him to write to the *Athenaeum*.[20] He praised that Journal for 'so steadily and courageously' opposing itself to 'those acts of barbarism which the modern architect, parson, and squire call 'restoration', and asked:

> Is it altogether too late to do something to save . . . whatever else of beautiful or historical is still left us on the sites of ancient buildings that we were once so famous for? Would it not be of some use once for all, and with the least possible delay, to set on foot an association for the purposes of watching over and protecting these relics, which, scanty as they are now become, are still wonderful treasures, all the more priceless in this age of the world, when the newly-invented study of living history is the chief joy of so many of our lives?

Morris followed up his letter by convening a meeting at his firm's premises in Oxford Street at which it was decided to form what became the Society for the Protection of Ancient Buildings (SPAB). The Society's first annual meeting was held on 21 June 1878, and among the original committee members were Carlyle, Ruskin, Burne-Jones, Holman Hunt, Alma Tadema, Mark Pattison, and Coventry Patmore. Morris himself was secretary. From this time on Morris & Co. ceased to accept commissions for stained-glass windows in ancient churches. The Society's influence seems to have gained ground steadily, and Morris himself found time for numerous site visits to deter proposed schemes of restoration (not always in the most tactful manner). It is interesting to note that Carlyle joined the Society mainly to affirm his opposition to the threatened demolition of some of Wren's city churches, which were not much appreciated at that time. Morris expressed somewhat inconsistent views on Wren's churches, although he wrote that there were 'unanswerable arguments for their preservation'.[21]

Morris set out the SPAB's approach in a founding Manifesto which partly reflected the views that Ruskin had expressed thirty years earlier. He urged those concerned for old buildings 'to put Protection in the place of Restoration, to stave off decay by daily care, to prop a perilous wall or mend a leaky roof by such means as are obviously meant for support or covering and show no pretence of other art, and otherwise to resist all tampering with either the fabric or ornament of the building as it stands'. Morris's Manifesto also expressed his view that if an ancient building had become inconvenient for its present use, rather than alter or enlarge it, it would be better to replace it with another building. Some fifty years later, in 1924, the SPAB issued an 'interpretative note' modifying that position:

> Where there is good reason for adding to an ancient building a modest addition is not opposed to the principles of the Society, provided (1) that the new work is in the natural manner of today, subordinate to the old, and not a reproduction of any past style, (2) that the addition be permanently required and will not in any sense be a building which future events will render inadequate or superfluous.[22]

In quoting the SPAB's view of 1924 we are getting ahead of our story, but it serves to show that the dogmatic tone adopted by Ruskin and Morris on this subject tended to persist, and we will find it appearing in other contexts later.

John Harvey reckoned that by the time Morris launched the SPAB 'the worst was already over'.[23] But the significance of the

SPAB's arrival on the scene can hardly be over-estimated. It was the first body specifically concerned with protecting ancient buildings. The Society of Antiquaries was primarily a learned society, and the Camden Society was concerned not with preserving what they found but with restoring it to what they conceived to be its original condition. The SPAB had a different agenda. It was also the first effective pressure group in the field of conservation, and the progenitor of the many such bodies that now crowd the conservation stage. Like many such descendants, it was single-minded and doctrinaire in pursuit of its beliefs. And it certainly had no interest in the advancement of modern architecture. Its influence over the whole course of the conservation movement had been profound. But their doctrinaire approach was redeemed by their scholarly expertise and their concern for the highest standards of craftsmanship in such restoration work as they could countenance.

A Dissenting Voice

Even in its early days, however, the SPAB attracted criticism from at least one source. This was the architect-critic Robert Kerr (born 1823 at Aberdeen), who kept up a sustained attack on the architectural establishment, and especially the RIBA, throughout his long life. He was one of the founders of the Architectural Association and in 1847 its first President. In 1884 he read a paper at the General Conference of Architects in London, on 'English Architecture Thirty Years Hence', which Pevsner reprints in full in *Some Architectural Writers of the Nineteenth Century*.[24] In the course of a vigorous polemic on the misguided architectural fashions of the past century he throws off a passing reference to the SPAB as that 'somewhat mysterious organisation'. He dismisses the fledgling society as follows:

> Its object is not even artistic, but historical: to preserve what is left of the past in the most indiscriminate way; whether good or bad, old or new, preserve it all, so that the reverie of the wayfarer may have not only something authentic, but everything veritable to dwell upon, even when the light of life, perhaps never a very bright light, has quite gone out.

Kerr was a maverick and consciously so. Like others of that breed, he stimulated his audiences without changing their opinions.

The SPAB's rigid adherence to its original concepts of preservation has attracted other critics. In 1972 John Harvey, an outstanding architectural conservationist himself, commented that 'It was a misfortune that his [Morris's] ephemeral attack upon the Restorers (who indeed deserved attack at that moment) should, by becoming in effect the Rules of a Society and thus fossilised in the hands of a perpetual trusteeship, have been elevated into an infallible Dogma.' Harvey, in attacking the SPAB, also gave an interesting assessment of the state of official and public attitudes towards preservation at that time, only twenty years ago, when he wrote that 'Much of the disrepute in which 'Preservationists' are held in some official quarters and by large bodies of the general public is directly due to the extreme form in which Morris's protest was cast.'

Nevertheless, the SPAB presented a formidable combination of consistent doctrine, scholarly resource and lobbying skills, which made it an increasingly potent force in the politics of conservation.

NOTES

1. Ross, M. (1996) *Planning and the Heritage*, 2nd ed. London: E. & F.N. Spon.

2. Boulting, N. (1971) The law's delays, in Fawcett, J. (ed.) *The Future of the Past*. London: Thames & Hudson, 1 pp. 11–13.

3. Clark, P. (1983) Visions of the urban community: Antiquarians and the English city before 1800, in Fraser, D. and Sutcliffe, A. (eds.) *The Pursuit of Urban History*. London: Edward Arnold.

4. Harvey, J. (1972) *Conservation in Buildings*. London: John Baker, p. 170.

5. Evans, J. (1956) *A History of the Society of Antiquaries*. Oxford: University Press.

6. This is such a memorable quotation that it is worth giving the exact source as cited by Pevsner: Webb, G. (ed.) *The Complete Works of Sir John Vanburgh: IV. The Letters*. London, 1927, p. 29.

7. Cornforth, J. (1993) The first Gothic chairs? *Country Life*, August 5.

8. Summerson, J. (1945) *Georgian London*. London: Penguin Books. Appendix 2 provides a very useful selective list of contemporary books about London architecture and topography under the Georges.

9. Kain, R. (1981) *Planning for Conservation*. London: Mansell.

10. Hunt, L. (1857) *The Old Court Suburb*., Vol. 1. I am indebted to my friend Hugh Ellis-Rees for this and the following item.

11. Croker, T.C. (1860) *A Walk from London to Fulham*. London: William Tegg.

12. Harvey, J. (1972) *Conservation of Buildings*. London: John Baker, p. 170.

13. Ferry, B. (1861) *Recollections of A. Welsby Pugin*. London, p. 85 (cited by Pevsner – see next item).

14. Pevsner, N. (1976) Scrape and anti-scrape, in Fawcett, J. (ed.) *The Future of the Past. Attitudes to Conservation 1740–1974*. Cambridge: Cambridge University Press, pp. 35–39.

15. Clark, K. (1974) *The Gothic Revival: An Essay in the History of Taste*, 4th ed. London: John Murray, p. 174.

16. Fawcett, J. (1976) A restoration tragedy: Cathedrals in the eighteenth and nineteenth centuries, in Fawcett, *op. cit.*, pp. 74–115.

17. Girouard, M. (1976) Living with the past, in Fawcett, *op. cit.*, pp. 116–139.

18. Rosenberg, J.D. (1964) *The Genius of John Ruskin: Selections from His Writings*. London: George Allen & Unwin.

19. Harvey (1972), *op. cit.*, p. 176.

20. Henderson, P. (1967) *William Morris: His Life, Work and Friends*. London: Thames and Hudson.

21. Harvey (1973), *op. cit.*, p. 176.

22. Dobby, A. (1978) *Conservation and Planning*. London: Hutchinson, Appendix 5.

23. Harvey (1972), *op. cit.*, p. 176.

24. Pevsner, N. (1974) *Some Architectural Writers of the Nineteenth Century*. Oxford: Clarendon Press.

3

Ancient Monuments

A Premature Initiative

In 1869 the First Commissioner of Works did a remarkable thing. At a time when the government had not as yet acknowledged any responsibility for ancient monuments, he asked the Society of Antiquaries to 'furnish a list of such Regal and other Historical Tombs or Monuments existing in Cathedrals, Churches and other Public Places and buildings as in their opinion it would be desirable to place under the protection and supervision of the government, with a view to their proper custody and protection.' The Society duly obliged, the Church authorities protested, and the government did nothing about it. Why the Commissioner took this bold initiative is not known, but it may well have had the untoward result of putting the Church authorities on their guard and prepared them to ensure that, when the Ancient Monuments Act of 1882 was introduced, ecclesiastical buildings in use were excluded from it. It was an unfortunate start to the government's involvement with ancient monuments and no doubt deterred it from going down that route for more than a decade.

The First Legislation

The Society of Antiquaries and the Society for the Protection of Ancient Buildings (SPAB) helped prepare the ground for the first legislative measure to do with conservation, but the credit for achieving this lies with one individual.

Sir John Lubbock was one of those present at the SPAB's first annual meeting in June 1878 and a member of the original committee. He was a man of little formal education (he left school at fourteen) but became a self-made scholar and a prominent public figure as Vice-Chancellor of London University and Chairman of the London County Council. He was a prolific writer on anthropology and archaeology, among other topics.

It was evidently as an archaeologist that his interest in preservation originated, and he was instrumental in saving Avebury stone circle from development: he bought the land that was threatened with development, and later bought the archaeological sites at Silbury Hill, West Kennet long barrow and Hackpen Hill. Purchase is the surest way to preservation, as Sir Arthur Evans found at Knossos.

Lubbock was the Liberal Member of Parliament for Maidstone. In 1873 he introduced his Private Member's *National Monuments Preservation Bill*. His proposals were far more comprehensive than those which the government finally adopted in 1882. There was to be a National Monuments Commission, including the Presidents of the Societies of Antiquaries of England, Scotland and London, the Keeper of British Antiquities at the British Museum,

Sir John Lubbock (Lord Avebury): progenitor of the first Ancient Monuments Act (1882). Contemporary cartoon by 'Figaro'.

the Master of the Rolls and other worthies (quite in the mode of the modern Quango). There would be a schedule of 'monuments' which, most significantly for future legislation on this subject, excluded inhabited houses and ecclesiastical buildings in use. It also excluded ruined buildings and any other remains or sites that formed part of a castle or abbey or stood in anyone's park or garden. In effect the Bill applied only to archaeological sites – mounds, tumuli, barrows, dykes etc., on farmland or in the open country. In the case of those sites that were not excluded but listed in the Schedule, the Commission could, after giving the owner notice (which was appealable to the courts), take a 'power of restraint'. If the owner then wanted to 'injure' the site he would have to notify the Commission, who would be required either to consent to the work or to purchase the site – for which Treasury approval would be required. The Commission would also have the power to acquire, by agreement with the owner, either the freehold or the right of public access, or to agree a power of restraint (comparable to the later concept of guardianship). Where it owned the freehold, the Commission would be enabled to inspect or excavate the site and to incur the cost of restoration or preservation (again with Treasury consent). Once vested in the Commission, sites would be derated.

One stands in admiration of Lubbock's legislative creativity. Here at once, as the result of one man's work and practical administrative ability, were all the essential components of what later became Britain's statutory system for the protection of ancient monuments. But it did not happen at once. Some of its key features did not reach the statute book for forty years.

Lubbock's Bill failed in 1873 and he reintroduced it without success in each of the following six years. It reached the stage of substantive debate in 1875. Lubbock tactfully explained that the Bill did not affect anyone's domestic property but concerned solely earthworks and similar sites 'which only wanted to be left alone'. But, as is so often the case, the Bill was loudly attacked not for what it proposed to do but for what it was assumed to do. One Member denounced it as 'legalising burglary by daylight' and authorizing 'acts of violence' by the Commission. So said Sir Charles Legard, a Tory baronet, who owned at least sixteen barrows himself. The Bill, he concluded, was 'an insult to the spirit and enterprise of private citizens who inherited these ancient monuments.' Lord Francis Hervey, a Tory lawyer, dismissed 'ancient monuments' as 'the absurd relics' of our 'barbarian predecessors'. Some Members spoke in support of the Bill, but the Secretary to the Treasury, Mr. W.H. Smith (son of the founder of that firm), replied for the government and expressed the Treasury's abhorrence of the potential expenditure. The Bill was lost that year and again in 1876 and 1877. Lord Percy seemed to reflect the majority view when he declared that the Bill proposed to take private property not for essential public purposes such as railways 'but for purposes of sentiment, and it was difficult to see where that would stop.' He had a point there.

Sir John Lubbock lost his seat in 1880 but returned the following year as Member for London University. The Liberals were now in power, and Gladstone agreed, since the government had no proposals of its own on the subject, to consider Lubbock's proposals. The result was the Act of 1882, but it was a poor creature compared with Lubbock's original proposals of 1873. It made a start on 'listing' with a Schedule of sixty-eight ancient monuments, comprising a miscellaneous collection ranging from Stonehenge to 'Hob Hurst's House and Hut' in Wiltshire. But there was now no element of compulsion and the Commissioners of Works would have no recourse against a recalcitrant owner. The Bill merely provided that the State could, if the owner agreed, purchase an ancient monument and maintain it; or, at the owner's instigation, could take it into 'guardianship' whereby the owner retained ownership but lost the right to demolish or interfere with it, while the State helped pay for maintaining it. These provisions related only to those ancient monuments listed in the Act or which were added to the list by Order in Council. Damage to a monument in guardianship incurred a fine of £5 or up to one month's imprisonment.

The Bill was still attacked in the Commons as an assault on private property 'to gratify the antiquarian tastes of the few at the public expense', but it finally received Royal Assent. Sir John Lubbock became Lord Avebury.

The historic significance of Lubbock's Act is that it was the first time that the national government was obliged to acknowledge any responsibility for conservation. Despite the growing concern for historic buildings and the formation of the SPAB, Lubbock was alone, at the start of his campaign, in calling on the government to get involved in the business of conservation. Even then the Act of 1882 was designed only to identify such monuments and to protect them from wilful or accidental destruction. It did not require the government to take any further action and there was as yet no sense of the need for a national policy for conservation. But, despite the Act's very modest start, the machinery was now in place and the way was open for the list of scheduled monuments to be steadily extended. Once set in motion, the engine of conservation could not be reversed but gradually gathered momentum.

A European Comparison

In 1905 a book was published, which is cited in all the bibliographies but perhaps little read. This is G. Baldwin Brown's *The Care of Ancient Monuments*.[1] Baldwin Brown was

Stonehenge, Wiltshire, one of the sixty-eight monuments scheduled by the 1882 Act: in 1836, water colour by John Constable RA; in 1900 before Ministry of Works partial reconstruction; and as it is today.

Professor of Fine Art at Edinburgh University. The scope of his short book was much wider than its title suggests. Its stated purpose was to describe 'measures in force for the safeguarding of ancient buildings and other objects of historical interest; for the maintenance of fitting aesthetic standards in the architecture of towns; and for the preservation of natural beauties of rural districts.' In practice, although the author took a catholic view of his subject, the main part of the book comprises a survey of the legislation and methods of historic preservation then in force in European countries (and some others). This fascinating and innovative account was based largely on reports from British Embassies, which resulted from an approach to the Foreign Secretary in 1897 by the Society of Antiquaries. Those reports had been published as a Parliamentary paper but, as Baldwin Brown remarked, the subject had excited little attention. His aim was to demonstrate that Britain, despite the Act of 1882, lagged sadly behind other European countries in its concern for historic buildings. He considered the 1882 Act to be an 'innocuous measure' that had had 'all its teeth drawn' in its passage through Parliament. He demonstrated that other countries had taken the lead in conservation at a far earlier stage. For example, Sweden in 1666, Portugal in 1721, Germany in 1780, Denmark in 1807, Greece in 1834.

John Harvey draws particular attention to the quite remarkable decree issued by Louis X,

Grand Duke of Hesse on 22 January 1818, which he prints in full together with the German original.[2] It is such a seminal document that it is well worth reproducing again here:

Louis, by the grace of God Grand Duke of Hesse and the Rhine, etc., etc.

Whereas the surviving monuments or architecture are among the most important and interesting evidences of history, in that from them may be inferred the former customs, culture, and civil condition of the nation, and therefore their preservation is greatly to be wished, we decree as follows:

1. Our Higher College of Building is charged with bringing all the discoverable remains of ancient architecture in the Grand Duchy of Hesse, which are worthy of preservation on historical or artistic grounds, into a correct inventory wherein their present condition is to be described and the ancient works of art to be found in them, such as paintings, statues and the like, are to be noted.

2. For the compilation of the historical part of this inventory the aforesaid College is to invite such as are most learned in the history of every province to collaborate in this patriotic purpose, and these are to communicate to that end the necessary information from the archives.

3. The most distinguished of these works, or the most ruinous, are to be completely and accurately surveyed and the drawings deposited with their description in our Museum.

4. Our Higher College of Building is hereby charged: to lay before us the inventory of these buildings considered worthy of preservation or illustration for our approval, in order to put in hand the preservation and repair of the same in conjunction with the various authorities; and to give us the appropriate advice thereon.

5. If it should seem necessary to put in hand alterations of one or other of these buildings, or to demolish one entirely, this is to be done only with the previous knowledge of the said College, and in appropriate cases after it has obtained a supreme approval.

6. If in the course of excavations or on other occasions antiquities are discovered, our officers are to take care that if possible they should be preserved, and notice thereof is to be given immediately to our Higher College of Building or to the Directorate of our Museum.

7. It is the duty of all public authorities to take the greatest possible care of the monuments made known in the aforesaid inventory, to which end the same shall be printed and communicated to them.

Darmstad, the 22nd of January 1818
LOUIS

There seems to be no doubt that this decree rapidly became known among incipient conservationists in Europe and, although it did not lead to much practical activity among the petty German states, it certainly prompted France and some other countries to get started. Britain, however, lagged behind and showed no greater inclination then than now to follow European initiatives. John Harvey, writing in 1972, caustically remarked that the 'brilliant intuition' which inspired the comprehensive but beautifully succinct (less than 400 words) decree of 1818 'puts to shame the long-winded legislation and regulations which beset the subject more than 150 years later.'[3]

As the nineteenth century progressed, more formal statutory provisions were introduced elsewhere in Europe. One of their striking features was that they were not confined to archaeological remains but in some cases extended protection to outstanding buildings of much more recent date: in Prussia up to about 1870; in Italy a limit of fifty years and in Hesse of thirty years. In Britain, by contrast, as Baldwin Brown remarked 'The shyness of the British parliamentary mind in the presence of monuments is so great, that in our Ancient Monuments Act of 1882 the legislature could hardly contemplate anything nearer in date than pre-historic standing stones and tumuli.'

Some countries had also realized that no protective system could be fully effective unless there was a method of recording or listing the buildings that ought to be preserved. Although France had originated such a system in 1837, it was criticized for being far too exclusive, since it was feared that a very restricted list would imply that buildings not included on it were of no importance. Dr. Gurlitt, who was in charge of listing in Saxony, remarked at an international congress at Lubeck in 1900 (perhaps the first of its kind) that 'When the scheduling goes no further than it does in France, it is of no use.' Romania had instituted a schedule that was updated every five years as more notable buildings were recorded, and in Canton Bern in Switzerland the list was updated every three years. Germany, however, was recognized to be far in the lead on listing: it was reported that the inventories compiled by the various German states already filled one hundred and thirty volumes, and that when complete the series would extend to two hundred volumes. Some European cities, including Vienna, Cologne, Lyons and Paris, had prepared their own lists (and, as we shall see later, the London County Council had started doing the same). Moreover, many countries had recognized the need for the State to intervene to protect, or if necessary to acquire compulsorily, monuments that were at risk of destruction.

Having reviewed all this activity elsewhere in Europe, Baldwin Brown concluded that 'in Britain, Government did less for monuments than is done in any other European country.' He noted that since the Act of 1882, only twenty-four of the sixty-eight monuments originally scheduled had been taken into guardianship, plus eighteen not in the original list; and that 'So far as any expenditure is concerned, these Acts have in Britain become almost a dead letter, while since the death in 1900 of the Inspector of Ancient Monuments, General Pitt Rivers, no successor has been appointed to the post.'

Baldwin Brown ended by urging the need for Britain to catch up with its European neighbours and, in particular, pressed for the setting up of a Royal Commission on Historical Monuments (RCHM) comparable to the Royal Commission on Historical Manuscripts that had existed since 1869 (this was a field in which Britain was acknowledged to be in the lead, but it raised less acute problems of invading the sanctity of private property).

Baldwin Brown's book, together with the continuing efforts of the Society of Antiquaries (who at that time were energetic and effective lobbyists), no doubt helped to bring about the setting up of the Royal Commission on Ancient and Historical Monuments (RCHM) in 1908 and the introduction of compulsory powers in the Act of 1913 (see below). He certainly deserves to be commemorated as one of the Founding Fathers of conservation in Britain.

More Ancient Monuments Acts

The Act of 1882 was amended in 1900 (again on Lubbock's initiative) to establish the right of public access to scheduled monuments and to enable county councils to take monuments into their care. The Act also reaffirmed that Scheduling could not apply to inhabited buildings. That restriction remained a key feature of the Ancient Monuments Acts for the next eighty years and was a major defect in Britain's statutory system of conservation. It prevented owners of buildings that were still inhabited from seeking financial aid from the government and it prevented the government from taking action to preserve a building that was falling down about the ears of its inhabitants. It also perpetuated a peculiarly antiquarian attitude towards ancient monuments, a distinctive culture among those responsible for them and something approaching disdain for other types of historic buildings that did not enjoy the protection of the Ancient Monuments Acts. This distinction between ancient monuments and other buildings of architectural or historic importance undoubtedly impeded the cause of conservation between the wars and was partly responsible for the loss of many important buildings. At a later date, as we shall see, that distinction became hopelessly blurred and added to the general confusion surrounding the subject.

In 1908 the Royal Commission on Ancient and Historical Monuments and Constructions of England was established (as Baldwin Brown had proposed) to 'make an inventory' of such monuments 'connected with and illustrative of the contemporary culture, civilization and conditions of life of people in England, excluding Monmouthshire, from the earliest times to the year 1700, and to specify those which seem most worthy of preservation.' The Commission published its first report, on Hertfordshire, in 1910. It produced an inventory of monuments as part of the report, and an accompanying volume of plans and detailed descriptions. In their first report they provided two separate lists, one identifying all the monuments that they felt were worth recording and a second shorter list nominating those which they considered worthy of preservation. This seems to have been the origin of the confusing distinction between archaeological remains that are officially recognised as such and the much smaller number that are scheduled ancient monuments. They made it clear that they saw their task as strictly one of compiling these lists and that they had no responsibility for intervening to secure their protection. They recommended that any such action should be the responsibility of 'a Government Department advised by an advisory body'. Such arrangements were established by the Act of 1913.

The Commission set itself the highest standards of scholarship and thoroughness but its very limited resources meant that the work proceeded at an infinitesimally slow rate: fifty years after its start only 20 per cent of the country had been covered and in 1963 the terminal date (for the age of buildings included) was abolished, thus ensuring that the Commission's work could continue to infinity. The Commission was very effective in compiling a detailed record of historic buildings, including many that have been lost over the past eighty years. While it played little part in preserving those not yet recorded, it is clear from its annual reports that it persistently urged the government to extend and strengthen the ancient monuments regime of the Act of 1913. The nature of the Commission's work, however, meant that it could not be used as the basis for the kind of comprehensive protection that was eventually adopted in the Planning Acts.

Ancient monuments continued to attract the attention of the legislators from time to time over the next eighty years. It would be tedious to trace them all in detail and only the few significant changes need be noted here. The objectives did not materially change: it has been a matter of progressively tightening the protective girdle. It is a curious feature that this legislative process was slow yet continuous, perhaps reflecting the characteristic pace of archaeological exploration and antiquarian study. A quite different tempo applied, and a different set of participants emerged, when attention turned to the protection of important buildings that were not ancient monuments.

In 1913 the Ancient Monuments Consolidation and Amendment Act[4] tidied up the earlier legislation, set up the Ancient Monuments Board to perform the advisory functions that the RCHM had abjured, and introduced new and important powers of compulsion. Owners had to be notified when a monument was scheduled and had to give one month's notice of any intention to alter or demolish. The Commissioners of Works were empowered to make a preservation order (a new concept), when recommended to do so by the Advisory Board, in respect of a scheduled ancient monument. Any such Order had to be confirmed by a Parliamentary Bill comparable to a Private Member's Bill and therefore subject to the owner's right to petition against the Bill.

The 1913 Act also gave the Commissioners the pre-emptive right to purchase when an ancient monument was offered for sale. The penalty for ignoring the Commissioners and selling privately was again set in the Bill at the princely sum of £5, but that was increased to £100 before the Bill was enacted (and, incredibly, that was still the penalty prescribed in the 1947 Planning Act; in 1991 the maximum penalty was increased to £20,000).

In addition to these new mandatory powers, the Act of 1913 imposed on the Commissioners of Works an important new duty. Section 12 provided that

> The Commissioners of Works shall from time to time cause to be prepared and published a list containing
>
> (*a*) Such monuments as are reported by the Ancient Monuments Board as being monuments the preservation of which is of national importance; and
>
> (*b*) Such other monuments as the Commissioners think ought to be included in the list.

The Act defined the scope of the task in very wide terms. The 'list' was to cover not only ancient monuments (which the RCHM was laboriously recording) but 'any other monuments or things (*sic*) which, in the opinion of the Commissioners of Works, are of like character . . . the preservation of which is a matter of public interest by reason of the historic, architectural, traditional, artistic, or archaeological interest attaching thereto; and the site of any such monument, or of any remains thereof.'

Thus the Commissioners of Works might appear to have been endowed in 1913 with the job of compiling the lists of buildings of architectural or historic interest which was given to the Minister of Town and Country Planning forty years later by the Town and Country Planning Act 1947.

The new Ancient Monuments Board, equipped with its own Inspectors, set about the task of compiling this list with rather more expedition than the RCHM, and by 1931 some 3,000 ancient monuments had been listed. Despite the apparently wide definition in the 1913 Act, however, the Board did not extend its listing activities beyond the traditional scope of the ancient monuments system. Buildings later than the sixteenth century, inhabited houses and ecclesiastical buildings in use were excluded, to be swept up later into the much more comprehensive lists prepared under the Planning Acts. It is curious that the Commissioners of Works, having been given the chance to establish themselves as the undisputed kings of listing and conservation, let the initiative pass to the infant Ministry of Town and Country Planning in 1947. One may infer that the Commissioners of Works, secure in their possession of the archaeological high ground, were inclined to look down on other types of historic buildings as a somewhat inferior species. If so, they were much mistaken in terms of emergent public opinion and the Whitehall power game.

The first preservation order under the 1913 Act was made in 1914 on an early Georgian house in Dean Street, Soho. The owner had restored it but could find no purchaser for it. The Act made no provision for financial assistance or compensation. The House of Lords upheld the owner's appeal and declined to confirm the order. After this set back, only two other (uncontested) orders were made under that Act.

While the Act of 1913 was on its way through Parliament voices were raised to urge the inclusion of other types of building, especially inhabited houses. A powerful exponent of this view was Lord Curzon (who owned three historic residences himself). He advocated the inclusion of mansions, manor houses 'and then, descending the scale, the smaller buildings, whether they be bridges, market crosses, cottages or even barns, which carry on their face the precious story of the past.' Curzon was supported by evidence given before the Select Committee by Charles Reed Peers, who had by now been appointed Inspector of Ancient Monuments. But it did not happen in 1913 and inhabited buildings were not brought into a statutory system of protection, with powers of compulsion, until some forty years later. When it did happen, they joined a different statutory system which was less secure but also far more comprehensive than the system that had evolved to protect ancient monuments.

Although some important buildings were lost as a result of the deliberate limitations of the Ancient Monuments Acts, it is probably just as well that the system devised to protect archaeological remains was not more widely extended. The curatorial and antiquarian culture of the ancient monuments system, with its emphasis on strict preservation, could not sensibly be stretched to cover the much wider range of conservation interests that developed later. Those interests involved a much more diverse set of motives and objectives. It is inconceivable that the stringent procedures designed to preserve ancient monuments in perpetuity could have been employed to protect over 500,000 listed buildings. The trouble is that it is now widely assumed that each listed building attracts – and deserves – the same protection as is afforded to those structures that happen to be defined as ancient monuments.

In 1931 the legislation was amended so that preservation orders did not have to be confirmed by Act of Parliament unless the owner objected, and in 1933 the procedure was changed again to enable objections to be heard at a public inquiry instead of by Parliamentary Select Committee. It also provided for compensation to be paid to owners adversely affected by a preservation order, which proved to be a deterrent rather than a help to local authorities in using those powers. The Act of 1933 completed, for practical purposes, the protective system for Ancient Monuments that had taken fifty years to evolve from the original Act of 1882.

There has been subsequent legislation ending with the comprehensive update comprised in the Ancient Monuments and Archaeological Areas Act 1979[5] but the later refinements (including provision for rescue archaeology) are not specially germane to our present purpose. For the present it need only be noted that the ancient monuments regime continued on its ponderous way, under the auspices of the Ministry of Works, with very little regard to the emergent planning system, although the shadows of inter-departmental jealousies can sometimes be seen lurking in the background (the 1933 Act laid down stringent requirements to ensure that the Minister of Health, who was responsible for town planning, consulted the Commissioners of Works, and the 1979 Act made it clear that where a building is both an ancient monument and a listed building, the former designation takes precedence).

Other Developments

Before turning to the Planning Acts we should first note some other relevant developments.

Many local archaeological societies were founded in the nineteenth century, for example Newcastle (1813), Oxford (1838), Wiltshire (1853), Cumberland and Westmoreland (1866), Leeds (1889). In 1875 was formed the Society for Photographing Relics of Old London, whose members were probably interested in old buildings more for their picturesque qualities than for their historic context, but they produced a valuable series of 120 photographs between 1875 and 1886, chiefly of seventeenth-century houses and inns, most of which had disappeared by 1900.

In 1894 Charles Robert Ashbee, a follower of Morris in the Arts and Crafts movement, set

The Clergy House (14th century), Alfriston, East Sussex. The first building acquired and restored by the National Trust (1896): before and after.

up a Committee for the Survey of the Memorials of Greater London. Its purpose was to compile a register of 'what still remains of beautiful or historic work in Greater London and to bring such influence to bear from time to time as shall save it from destruction or lead to its utilisation for public purposes.' In December 1896 the LCC convened an important conference of fifteen interested bodies to discuss the need to institute an official inventory of historic buildings in London. This led to the Survey being carried forward in partnership with the London County Council. It has resulted in a monumental series of some fifty volumes, which is still being carried forward parish by parish.[6] When the GLC was abolished in 1986 the Survey was incorporated into the Royal Commission on the Historical Monuments of England. In 1898 the LCC obtained statutory powers to buy historic buildings or to provide funds for their maintenance. The first use that they made of their new powers was to buy 17 Fleet Street, one of the finest half-timbered buildings left in London. At about this time the LCC also took over from the Society of Arts what became the 'Blue Plaque' scheme for identifying buildings associated with historical events or famous people. Some other County Councils (notably Northamptonshire) also became active in recording historical buildings, and the Victoria County Histories began publication (originally more than 150 volumes were contemplated).

In 1895 the National Trust was founded: its full name was the National Trust for Places of Historic Interest or Natural Beauty.[7] For the first fifty years or so of its existence the Trust concentrated chiefly on acquiring and protecting beautiful landscapes. But from its earliest days it was also concerned with ancient buildings. In 1894 Octavia Hill, one of the three founders of the Trust, moving the resolution that led to the setting up of the Trust, said that its purpose was to 'save many a lovely view or old ruin or manor house from destruction and for the everlasting delight of thousands of the people of these islands.' The first building acquired by the Trust was the half-timbered (and ruinous) Clergy House of Alfriston in Sussex. It was brought to the Trust's notice by the SPAB and bought from the Ecclesiastical Commissioners in 1896 for £10. An appeal raised £350 for repairs. In 1897 the Trust acquired the fourteenth-century Old Post Office at Tintagel for £200 and then raised a loan for the acquisition of the sixteenth-century Joiners' Hall in Salisbury. Meanwhile it had begun to acquire various small parcels of open land in Cornwall, the Lake District and the Fens.[8] During the first twenty-five years of its existence the Trust acquired over eighty properties, most of which were open spaces and coastline but also included two manor houses, a dozen smaller historic buildings, two dovecotes and a series of bridges over the River Wey.[9] Thus began the Trust's great work. A full account of it is given in the volume published to mark its centenary in 1995 *From Acorn to Oak Tree*.[10]

In the 1920s the Royal Society of Arts pursued a campaign to save old cottages and in 1929 secured the preservation of the whole village of West Wycombe in Berkshire by acquiring the freeholds (later transferred to the National Trust). But this seems to have been a temporary diversion from the Society's traditional concerns and its growing role in wider fields of education and training. It did not play a leading role in conservation in later years, when a more diverse cast of players has emerged.

In 1924 the Royal Fine Art Commission was established to advise the government and other public bodies on matters of artistic importance, including conservation. The Commission occasionally lent its weight to the cause of preserving particular buildings but its interventions were infrequent and unpredictable. It preferred to operate by influence and covertly. It is only in recent years that it has 'come out' and begun to express its views publicly and forcibly on a wide range of architectural issues including conservation.

Against this background we can turn next to the Planning Acts and their place in the policy history of the heritage.

NOTES

1. Brown, G.B. (1905) *The Care of Ancient Monuments*. Cambridge: Cambridge University Press.
2. Harvey, J. (1972) *Conservation in Buildings*. London: John Baker, pp. 27–28 and Appendix 1, pp. 208–209.
3. Harvey (1972), *op. cit.*, p. 28.
4. 3 and 4 Geo. V c.24.
5. Eliz 2 c.46.
6. Hobhouse, H. (1994) *London Survey'd, The Work of the Survey of London 1894–1994*. London: RCHM.
7. Waterson, M. (1994) *The National Trust: the First Hundred Years*. London: BBC Publications.
8. Murphy, G. (1987) *Founders of the National Trust*. London: Christopher Helm.
9. Murphy (1987), *op. cit.*, p. 121.
10. Jenkins, J. and James, P. (1994) *From Acorn to Oak Tree the Growth of the National Trust 1895–1994*. London: Macmillan. For a more critical view of the Trust's activities see Weideger, P. (1994) *Gilding the Acorn: Behind the Facade of the National Trust*, London: Simon and Schuster.

4

PLANNING

While the statutory system for ancient monuments was taking shape from 1882 to 1933, the concepts of town and country planning were also evolving – and, of course, it is possible to trace the origins of town planning back to very early times and ancient civilizations.[1] But in the modern sense the idea of town and country planning evolved rapidly in the last twenty years of the nineteenth century, largely as an off-shoot from the Victorian concern with public health and housing, but acquiring a wider and more distinctive role from the writings of Ebenezer Howard, Geddes and others. This is not the place to trace that history. It has been well dealt with in, for example, William Ashworth's *The Genesis of Modern British Town Planning: A Study in Economic and Social History of the Nineteenth and Twentieth Centuries*.[2] Our interest is in how the subject of conservation found its way into the planning system. As we have seen, there had been some pressure to extend the legislation dealing with ancient monuments so as to include inhabited buildings and those of later date. But it seems that those responsible for the ancient monuments system did not have any such empire-building ambitions. It was left to the *arriviste* planners to take such buildings under their wing but, having done so, they took a long time to do anything much about them.

HOUSING, TOWN PLANNING ETC. ACT 1909

It was the Housing, Town Planning Etc. Act of 1909[3] that first introduced the concept of Town Planning Schemes, the precursors of today's development plans.[4] The Fourth Schedule contained a list of matters to be included in such schemes and which were to be dealt with in more detail by General Provisions prescribed by the Local Government Board. The fourth of these items was 'The preservation of objects of historical interest or natural beauty'. Those words were not in the Bill as introduced by the government. They were added at the Committee stage by an amendment moved by Mr. Morrell on 3 December 1908. Philip Edward Morrell was the liberal member for South Oxfordshire (1906–1910) and for Burnley (1910–1918). He was the husband of Lady Ottoline Morrell, the famous Edwardian hostess and literary figure, who later held court at Garsington Manor in Oxfordshire. The *Dictionary of National Biography* notes that he 'shared her aesthetic tastes and advanced views'. Certainly they would have known of William Morris and the SPAB, but I have not found direct evidence of a closer connection that would account for his Parliamentary initiative in the Act of 1909.

Mr. Morrell thus joins Sir John Lubbock as one of the Parliamentary originators of conservation. His amendment was not discussed in Committee or at later stages in the Bill's progress: so objects of historical interest made their way into the planning system by stealth. This is a little disappointing for our present purpose, since debates on the Bill might have been expected to throw some light on the sources of that historic amendment. There was more interest shown in the Bill's provisions concerning the preservation of commons than in that relating to historic buildings.

The Local Government Board (LGB) promptly issued a circular and memorandum on the 1909 Act, followed by procedural regulations.[5] But they baulked at providing any guidance on the 'matters to be dealt with by General Provisions' in the Fourth Schedule, including the preservation of objects of historical interest. The Act also enabled the LGB to prescribe separate sets of provisions adapted for 'areas of any special character'. The memorandum announced that 'The Board are giving consideration to the subject of General Provisions, but they will not be able to issue them immediately.' In fact they never did so, either in general or for areas of special character. Local authorities were left free to get on with preparing their town planning schemes as best they could.

The first of these schemes were two for Birmingham in 1913 and one for Ruislip-Northwood in 1914 (which has come to be regarded as the classic example, partly because it included the first attempt at aesthetic control). None of these schemes had any proposals relating to objects of historical interest, nor have I found any in other early schemes. That is hardly surprising since these pioneering exercises in statutory town planning were focused entirely on the laying out of new areas for development.

In 1914 the Local Government Board produced revised procedure regulations under the Act of 1909. Among the items of information to be furnished to the Board in connection with any application for approval to prepare a town planning scheme was

> (*j*) Information as to any monuments or ancient monuments, within the meaning of the Ancient Monuments Consolidation and Amendment Act, 1913, situate within the area included in the scheme, and as to the manner in which they would be affected.

Thus in 1914 the legislation concerning ancient monuments is brought into contact for the first time with the upstart town planning legislation. Nothing of interest appears to have resulted from this brief encounter.

Housing and Town Planning Act 1919

No relevant changes were made in the Housing and Town Planning Act 1919,[7] although that Act was notable for making the preparation of town planning schemes obligatory for all boroughs and county districts with a population of more than 20,000. The Ministry of Health (which had succeeded the Local Government Board in 1931) issued a Model Form of Preliminary Statement, which went into considerable detail about the form and content of schemes, but it contained no reference to historic buildings or 'areas of any special character'. The emphasis was still exclusively on 'areas to be developed'.

Housing Etc. Act 1923

Then came the Housing Etc. Act 1923.[8] Note that 'Town Planning' has now been absorbed in the 'Etc.', which must strike us as rather demeaning. It probably also accounts for the fact that none of the standard works on the history of town planning refers to the Act of 1923. But it is highly relevant to our subject on account of Section 23, which reads as follows (italics added):

Power to make town planning schemes in special cases.
– Where it appears to the Minister that on account of *the special architectural, historic or artistic interest attaching to a locality it is expedient that with a view to preserving the existing character and to protect the existing features of the locality* a town planning scheme should be made with respect to any area comprising that locality, the Minister may, *notwithstanding that the land or any part thereof is already developed*, authorise a town planning scheme to be made with respect to that area prescribing the space about buildings, or limiting the number of buildings to be erected, or prescribing the height or character of buildings, and, subject as aforesaid, the Town Planning Acts, 1909 to 1923 shall apply accordingly.

This section clearly marks a new phase in our history. For the first time there is reference to 'architectural' as well as 'historic' interest – and also to 'artistic' interest, a term which was dropped in the 1947 Act (it is not clear whether it referred to scenes of artistic appeal or to scenes that were of interest because of artistic associations – for example the setting of Constable's *Flatford Mill* which might qualify on both counts). Secondly, the section refers to localities and areas rather than to individual buildings. Thirdly, it envisages what we would now call 'aesthetic' control designed to preserve 'the existing character' of the locality, including controls prescribing the 'character' of new development. Thus the 1923 Act in effect introduced the concept of 'conservation areas', although this term does not make its appearance in legislation until 1967. Despite the novelty of this new provision, no reference was made to it during the Parliamentary debates on the 1923 Act, when attention focused almost entirely on the housing provisions.

Following the new Act, the Ministry of Health published *Model Clauses for Use in the Preparation of Schemes* but these contained no reference to areas of special architectural, historic or artistic interest. In 1925 there was a consolidation Act incorporating new powers (notably the remarkable Section 16 enabling local authorities or non-profit-making bodies to acquire land compulsorily 'for the purpose of garden cities'); but nothing on historic buildings, apart from repeating the Fourth (now First) Schedule of the 1909 Act.

The apparent lack of interest by the Ministry in these various provisions concerning historic buildings and areas of special architectural interest in the Acts of 1909 to 1925, is puzzling. As they were not touched on in debates on the Bills, and led to no official guidance, one wonders how they got into the Acts in the first place. It almost seems as though there were a mole in the Ministry of Health, an early conservationist perhaps, who contrived (with Mr. Morrell's assistance) to insert these provisions surreptitiously. If so, his efforts did not lead to anything beyond the legislation.

Town and Country Planning Act 1932

The 1925 Act had been the first Act in which 'Town Planning' was not in legislative harness with 'Housing'. That Act was superseded by the Town and Country Planning Act 1932 when for the first time 'Country' joined 'Town' and thus affirmed the comprehensive scope of modern town and country planning.[9] The bill was first introduced by the Labour government in 1931 but failed to get beyond the committee stage. The new National government re-introduced it in 1932. It commanded a wide measure of bipartisan support, although some strident voices were raised against it as a 'Socialistic' measure. One Member condemned it as 'bureaucracy run mad'.

The 1932 Act was a remarkable piece of legislation and not sufficiently celebrated in the history of town planning. It began with a new and comprehensive statement of the scope of planning schemes (italics added):

> A scheme may be made under this Act with respect to any land, whether there are or are not buildings thereon, with the general object of controlling the development of the land comprised in the area to which the scheme applies, of securing proper sanitary conditions, amenity and convenience, *and of preserving existing buildings or other objects of architectural, historic or artistic interest* and places of natural interest or beauty, and generally protecting existing amenities whether in urban or rural portions of the area.

This definition was perhaps less satisfactory than that in Section 23 of the 1923 Act (quoted above). It dropped the references to 'locality', 'character' and 'existing features', and seems to focus more narrowly on 'buildings' and 'objects', but it did introduce the term 'amenity', which was to become one of planning's main props. The 1932 Act has, of course, been totally overshadowed by the Act of 1947, which is generally assumed to be the foundation stone of the British planning system. It is certainly true that the 1932 Act's provisions on compensation and betterment proved to be an inadequate basis for effective planning and were superseded by the radical settlement of 1947. But as regards the machinery of the planning system, the 1932 Act anticipated most of the features of the 1947 Act and in some respects went further. In addition to the by then familiar system of planning schemes and development control, it provided for General Development Orders, specific powers to regulate 'the design or external appearance of buildings', powers of enforcement, planning agreements, building and tree preservation orders, and control of advertisements. In addition it envisaged arrangements for regional planning among groups of local authorities.

So far as our present subject is concerned, however, the 1932 Act contained nothing new other than the provision for building preservation orders, which extended to other types of building (notably occupied houses) the powers of compulsory preservation that had existed for ancient monuments since 1913. As regards conservation, however, the 1932 Act was fatally flawed in three respects.

The first defect, as already noted, was that the compensation structure meant that owners would have to be fully recompensed for any loss of value resulting from a building preservation order. This deterred local authorities from using those powers.

The second defect was that while the Act made specific reference to the protection of buildings of special architectural or historic interest, it provided no method of identifying or listing them (as noted above, the LCC was already supporting the Survey of London, but this was a long-term scholarly undertaking not an instrument of administration). Presumably it was thought that local authorities and the Minister would use their judgement in deciding which buildings merited that description when preparing a town planning scheme or approving a building preservation order – or possibly it was thought that the Commissioners of Works would rapidly complete their list under Section 22 of the Act of 1913 (as noted above). It did not seem to occur to those concerned that, without some form of prior listing, important buildings could be lost to demolition due to the ignorance or cupidity of their owners and before a preservation order could take effect.

This deficiency was not dealt with until the Acts of 1944 and 1947. But the point was raised in an entertaining episode during the committee stage of the Bill in April 1932, when two scions of the Salisbury family, then in the House of Commons, addressed their aristocratic minds to it. Viscount Cranborne urged that the new power in Clause 17 to initiate building preservation orders should not be left to local authorities, since 'One of the chief failings of local authorities is a complete lack of artistic sensibility.' He proposed that the Minister should also have that power and that he should be advised by an expert advisory committee, as proposed by the Member for Salisbury, Major Despencer-Robertson. That might have been a good idea, but Lord Cranborne had apparently assumed that the purpose of Clause 17 was not simply to enable preservation orders to be served on individual buildings but required the preparation of a map identifying all the historic buildings in an area. Again, this might have been a good idea but it was not what the Bill proposed. Lord Cranborne saw no difficulty in the advisory committee producing such maps. 'There is no lack of experts in this country – innumerable people who have not very much to do, and plenty of taste, who would be able to do this work perfectly well.' Moreover, he did not see it as a burdensome task: 'All they would have to do would be to take a large scale Ordnance Survey map and mark the buildings which ought to be preserved. There would not be a day's work in it.'

This sanguine assessment prompted his noble kinsman, the Marquis of Hartington, to respond. He announced that he was a member of the Royal Commission on Ancient and Historical Monuments, and that the Commission reckoned that it would take 86 years to complete their survey. Viscount Cranborne in reply to this said that 'the truth about him is that he is living in a period of about 50 years ago'. However, he withdrew his proposal on being assured by the Minister (Sir Hilton Young – father of Wayland Kennet, see chapter 12) that an advisory committee was unnecessary since the Bill required him to consult the Commissioners of Works on any proposed preservation order. Thus the idea of having an advisory committee to draw up maps or lists identifying the historic buildings worthy of preservation in any area got lost in the parliamentary confusion.

The third flaw in the 1932 Act resulted from a botched compromise that emerged when the Bill was discussed at committee stage in the House of Lords. This concerned the provisions in Clause 17 for building preservation orders. This clause had already been attacked in the Commons by a small group of Members led by Mr. Beaumont, MP for Aylesbury. He declared that 'it seems to me that the principle of the clause is unnecessary and vicious . . . undesirable and unwise.' He invoked the venerable image of the Englishman's house as his castle, and protested that the clause would make it 'a local authority's playground'. At that stage the government successfully resisted the attempt to remove the clause from the Bill, but when it reached the House of Lords the attack was renewed. Lord Banbury of Tougham sought to delete the offending clause, while Lord Cranworth proposed amending it so as to prevent demolition but allow alterations. The government spokesman initially resisted that proposal on the grounds that it would introduce an undesirable ambiguity in distinguishing between alteration and demolition. But in the event the government accepted the amendment and it was passed without comment when the Bill returned to the Commons.

The reason for the government's surrender on Clause 17 was partly that they were concerned to achieve the Bill's more important objectives and, as often happens, were prepared to make concessions on what appeared to be relatively minor matters. It seems also that doubts had arisen in official minds about the distinction between demolition and alteration: so it seemed simplest to confine the clause to demolition. But in practice, of course, almost anything short of complete demolition and removal of a whole building could be claimed to be 'alteration'. The only way to overcome that ambiguity was not by confining the new powers to demolition but by ensuring that they covered both demolition *and* alteration. That defect was not put right until 1947.

Thus the 1932 Act hardly advanced the cause of conservation. It provided no mechanisms for identifying or listing historic buildings; the new powers to make building preservation orders were defective; and the compensation provisions were a deterrent to using them. In the event no such preservation orders were ever successfully made. One or two draft orders were begun but never reached approval. The London County Council relied on its own Local Act powers and used them successfully in a few cases. But overall, despite the successive legislative provisions concerning historic buildings from 1909 to 1932, there was still no comprehensive policy or machinery for identifying and protecting them. Ancient monuments had their own regime, although the antiquarian approach to that subject meant that the protective net that was supposed to safeguard them was extended only very gradually.

The truth is that the need for a comprehensive system of conservation was not yet generally acknowledged. During the debates of Clause 17 of the 1932 Act nobody protested when the Marquis of Hartington (who was a member of the Royal Commission) called it 'an abominably bad clause', and asserted that 'These buildings have been preserved to us not by Acts of Parliament but by the loving care of generations of free Englishmen who did not know what a Minister of Health was and who also did not know, in many cases, what a District Council was.'

One further point to note on the 1932 Act is that it required the Minister of Health to consult the Commissioners of Works before approving any planning scheme that involved 'the removal, pulling down, or alteration of a building of special architectural or historic interest', and also before deciding whether to approve a building preservation order. For good measure, the Act also specifically provided that none if its provisions should affect any of the powers of the Commissioners of Works under the Ancient Monuments Acts. Evidently an interdepartmental row had been brewing as to who was to be responsible for buildings of special architectural and historic interest; but they were now locked into the Planning Acts rather than being treated as an appendage to Ancient Monuments legislation. This rather odd separation of responsibilities prevailed until 1971, as we shall see later. In reality, one can infer from the annual reports of the Royal Commission and the Commissioners of Works that those custodians were not so much anxious to take responsibility for an unquantified mass of lesser buildings as to retain their grip on the authenticated stock of ancient monuments.

NOTES

1. Kostof, S.(1991) *The City Shaped: Urban Patterns and Meanings through History*. London: Thames and Hudson and Kostof, S. (1992) *The City Assembled: the Elements of Urban Form through History*. London, Thames and Hudson.
2. Ashworth, W. (1954) *The Genesis of British Town Planning*. London: Routledge and Kegan Paul. Cullingworth, J.B. (1993) *Town and Country Planning in Britain*, 11th ed. London: Unwin Hyman. Cherry, G.E. (1974) *The Evolution of British Town Planning*, London: Leonard Hill.
3. 9 Edw. 7 c.44. A.R. Sutcliffe gives an interesting account of the evolution of the 1909 Act in *Town Planning Review*, **59**(3), July 1988, but this does not deal specifically with Morrell's amendment that introduced the reference to 'the preservation of objects of historical interest' into the Fourth Schedule.
4. I am grateful to Simon Morley for letting me see his unpublished MA thesis for Birmingham University Faculty of Law,Major Aspects of the Development of Statutory Town Planning 1909–1932 (this does not deal specifically with conservation but recounts the early legislative history of town planning).

5. Circular No. 17, 31 December 1909 reprinted in Local Government Board Report 1909–10. Part II p. lvii. Memorandum published separately. Town Planning Procedure Regulations SR&O 1910 No. 436.

6. Most of the memoranda etc. on this subject issued by the Local Government Board and the Ministry of Health can be identified only by their title and date: there was no consistent series.

7. 9 and 10 Geo 5 c.5.

8. 13 and 14 Geo 5 c.24.

9. 22 and 23 Geo 5 c.48.

5

Demolition and Inaction 1900–1940

At this point we should pause in tracing the legislative history and assess what had been happening to the country's stock of historic buildings during the first forty years of the twentieth century and whether there had been any signs of growing public concern for their conservation. Something must surely have happened that prepared the way for the new regime ushered in by the Planning Acts of 1944 and 1947.

Mediaeval and Tudor Buildings

We have already seen how in the nineteenth century the activities of the Victorian church restorers led to the setting up of the SPAB, which was concerned not only with ecclesiastical buildings but with other types of mediaeval buildings that displayed traditional craftsmanship. But despite the widespread support that the SPAB attracted from the artistic and literary establishment, and its occasional successful rescue operations, it had little or no effect in stemming the vigorous activities of Victorian commercial developers. Almost all vestiges of mediaeval and Tudor buildings in the City of London, and in many provincial cities, were swept away. And that is not surprising, given the state of dilapidation and disease-ridden poverty that prevailed in those districts. The public health reformers added their zeal to the developers' entrepreneurial propensities, and civic pride drove a desire for metropolitan improvements that cleared away much of the past.

It is hardly worth regretting the loss of that tattered mediaeval fabric. It was inevitable, and the buildings that replaced it were often of a quality and architectural richness that has led to demands for these in their turn to be preserved. But it might be thought that the successive legislative provisions from 1909 to 1932 reflected a growing awareness of the built heritage and a more widespread concern for its conservation. In fact there is little such evidence in the first thirty years or so of the century. It remained very much a minority interest and continued to be so, even after the 1947 Act provided effective machinery for the purpose.

In retrospect, however, one can see how the continuing loss of outstanding historic buildings must have had a cumulative effect that led to growing concern among architectural scholars and historians. The general climate of opinion, however, was shaped not so much by the loss of individual buildings as by reaction against the impact on the countryside of the rapid spread of suburbia and ribbon development in the 1920s and 1930s, which was seen to be engulfing historic towns and unspoilt villages. This concern for the countryside and rural England swept up the more specialist concern for historic buildings and was carried forward into the post-war planning legislation. That climate of opinion was fuelled by Clough William-Ellis's famous diatribe of 1928, *England and the Octopus*.[1]

GEORGIAN AND VICTORIAN BUILDINGS

The student of conservation must be astonished by the reckless way in which even the most distinguished buildings were thrown away in the first forty years of this century. Among the most notable losses were the redevelopment of Nash's Regent Street by the Crown Commissioners in the 1920s; the south front of Adam's Adelpi in 1936; and most of Soane's Bank of England rebuilt in 1925–30. Many of the great aristocratic town houses dating from the

The Piccadilly Hotel, Regent Street, London, by Norman Shaw (1905) begins the redevelopment of Nash's Regent Street.

The Adelphi, London, by Robert Adam (1768). South front before (1932) and during demolition (1936), replaced by offices.

eighteenth and nineteenth centuries surrendered to the combined pressures of an obsolescent life-style and the irresistible development value of their sites: Dorchester House, the most magnificent, in 1924; Norfolk House in St James's Square in 1936; Chesterfield House in 1937; Lansdowne House in 1936 (20 feet chopped off the front for road widening and much of the interiors sold off); and many others. Christopher Sykes in *Private Palaces* lists sixty-five London mansions, mainly in the vicinity of Mayfair, Park Lane and Piccadilly, of which over half have disappeared.[2] Georgian Bloomsbury started to give way to new buildings for the University of London (a process that continued into the 1970s). Many prominent Victorian public buildings were demolished and replaced, often by institutional or government offices, and these were seldom mourned at that time when Victorian architecture had few admirers.

Northumberland House

A striking example of non-preservation was Northumberland House, a huge seventeenth-century mansion with about 150 rooms and a large garden that at one time extended to a

Norfolk House, St James Square, London (1756): the ballroom. Demolished 1956 and replaced by offices

watergate on the Thames.[3] It was the last of the Thames-side palaces and the London home of the Dukes of Northumberland. In the 1870s the Metropolitan Board of Works decided that it had to be demolished to cut a new road from Trafalgar Square to the Embankment. There was some antiquarian opposition but no great public protest. The Duke expressed concern at the loss of his ancestors' house but reconciled himself to £500,000 compensation. The distinguished architect Sir James Pennethorne produced a perfectly plausible plan to divert the new road around Northumberland House and most of the garden. But it was demolished in 1876 to make way for Northumberland Avenue. The Grand Hotel replaced it and opened in 1880. About a hundred years later the Grand Hotel was demolished, after a great deal more fuss than its predecessor generated, to be replaced by an otiose replica in the form of an office block – the result of an architectural competition and the timidity of the planning authority. It is now called Grand Central Buildings.

London Churches

The most astonishing illustration of the destruction of historic buildings and the lack of any system of preservation, is the story of the City churches. Ecclesiastical buildings had always been excluded from the Ancient Monuments Acts and were exempt from preservation orders. The problem of redundant churches in the City of London was first raised in the 1850s when the Bishop of London drew up a list of twenty-nine churches which he proposed to demolish, including several by Christopher Wren and some mediaeval churches that had survived the Great Fire. This did prompt a public outcry; *The Times* called it a 'vast act of desecration'. But an Act of Parliament was obtained which led to the destruction of twenty-two churches. About eighty years later the Church authorities launched a further assault. The Phillimore Report in 1919 recommended the demolition of another nineteen churches (including several by Wren and others by Hawksmoor and Dance). A Bill to facilitate this was brought before Parliament in 1926 but was eventually defeated after widespread public protest and the opposition of the City Corporation. The amazing thing is not that the Bill was lost but that the Diocese of London brought it forward in the first place. Sadly, several of the churches then reprieved were lost in the blitz. The question of redundant churches is dealt with more fully in chapter 16.

Waterloo Bridge

One other case which caused a major rumpus between the wars was the demolition of Waterloo Bridge. This was particularly notable because for the first time a powerful consortium was brought together to oppose the proposal. The RIBA, the Town Planning Institute, the SPAB and the Royal Academy combined to put forward alternative schemes to save the bridge, which had been designed by John Rennie the elder and constructed by his son between 1811 and 1817. It was sad that the decision to rebuild the bridge had to be taken by the LCC, who had a good record in conservation. Parliament refused to approve a Bill to finance the work, but in the end the LCC decided to fund it themselves and the bridge was closed for demolition in 1934. At least London got a well designed modern bridge in its place: the Rennies might not have been too dismayed. The significance of the Waterloo Bridge story is not that the old bridge was demolished but that for the first time the preservationist opposition was well-orchestrated, professional and determined. It set a precedent for future battles.

A Resounding Protest

While the likes of Osbert Sitwell and other literati expressed occasional disdain for the depredations of developers and the cupidity of wealthy freeholders, one voice was raised in vituperative protest. This was that of Robert Byron, whose precocious talent flared briefly in the 1930s before his death in the war. As a young man he had already published several books on art and architecture, and in 1937 his classic travel book *The Road to Oxiana*. In that year also he published in the *Architectural Review* his article *How We Celebrate the Coronation*.[4] He reviewed what had been happening to London's architectural heritage and lashed out at those he held responsible – 'the leaches of Whitehall, the spiders of the Church, the long-nosed vampires of high finance and the desperate avarice of the hereditary landlords.' He declared

> Today architecture, as controlled by speculators and officials, is a forgotten art: when a work of genius or a building of famous associations is demolished, there may be compensation for the landlord, but there is none – in this heyday of democracy – for the public.

He then listed ten of the most conspicuous buildings lost in the past few years including (in addition to those already mentioned above),

Carlton House Terrace, The Mall, London, by John Nash (1827). The Crown Estate Commissioners proposed to replace it by a department store in the 1930s but it survived this and a scheme to add several storeys of offices in 1946.

Sir Joseph Bank's fine house (by Adam) in Soho Square; Sir Joshua Reynold's house in Leicester Square (demolished to make way for an extension to the Automobile Association's offices); Pembroke House, Whitehall (part of the site on which the Ministry of Defence now stands); and All Hallows, Lombard Street, which the Church authorities – 'these befrocked and dog-collared vandals' – intended to demolish. Most astonishing of all was the Crown Estate Commissioners' plan in 1932 to demolish Carlton House Terrace for redevelopment. It was envisaged that it should be replaced by a new building for Swan and Edgar, the department store. That outrageous proposal was eventually abandoned, but not before the Commissioners had allowed No. 4 Carlton Gardens (an essential part of Nash's composition) to be torn down and replaced by a block of offices in a grotesque pseudo-Nash style (in stone not stucco) with double the number of storeys. That house had been the home of two Prime Ministers – Lord Palmerston and Arthur Balfour. The new building has a fatuous plaque on it:

LORD PALMERSTON
1784–1865 STATESMAN LIVED HERE
TABLET FIXED 1907
PREMISES REBUILT 1933
TABLET REFIXED 1936

Having lambasted the perpetrators of destruction, Byron then mounted a vigorous attack on the ancient monuments regime, taking up where Robert Kerr had left off. He was contemptuous of the 'feeble casuistry of the Ancient Monuments Acts that exclude all architecture' – referring of course, to the fact that those Acts excluded ecclesiastical buildings in use and all inhabited houses. He concluded that the Acts were 'designed to preserve precisely those structures which, as examples of the country's art and genius, are the least worth preserving because they are in worst condition'.

Rather surprisingly, the *Architectural Review* offered no editorial support for Byron's fulminations: the Editor, the youthful J.M. Richards, devoted his editorial that month to the M.A.R.S group's landmark exhibition of the work of the modern movement at the New Burlington Galleries. But Byron's article was republished as a pamphlet and undoubtedly made a profound impact on those who cared about these things. He showed that one did not have to be polite or respectful in dealing with this subject, and, although it may have achieved little immediate result, it seems certain that it influenced those who took up the cause in the 1940s. For that he deserves to be commemorated as the first modern apostle of conservation.

THE GEORGIAN GROUP

It is not necessary to catalogue the many other fine Georgian and Victorian buildings that were discarded in the latter part of the nineteenth and early twentieth century. Such lists make depressing reading, but they are sadly chronicled and illustrated in Hermione Hobhouse's *Lost London*,[5] John Summerson's *Georgian London*,[6] and Anna Sproule's *Lost Houses of Britain*,[7] which takes the tale beyond London and into the great rural estates and their prodigious mansions.

Despite these drastic losses, there was not much evidence of sustained or effective public opposition to what was happening. There were the occasional expressions of regret and nostalgia in editorial columns, but a sense of resignation rather than sustained protest.

Against the forces of commercial gain and public apathy, the Georgian Group was formed in 1937. It originated as a minority group within the SPAB founded by those who felt that the Society was too little concerned

Devonshire House, Piccadilly, London by William Kent (1733). The ballroom. Demolished 1924 and replaced by offices (opposite).

with buildings later than the seventeenth century. The leading lights were Robert Byron and the architect Lord Esher. It was probably the Georgian Group, more than any other conservationist body, that led the way to the new regime established by the Planning Acts of the 1940s. The Group gathered together a formidable band of architectural scholars and practising architects. Moreover they brought to conservation something different from the antiquarian interests of the SPAB. They were concerned with historic buildings not chiefly as historical relics but as architecture and as vital components of the urban scene. They brought a new set of aesthetic values to bear, backed up by no lesser standards of scholarship and discrimination. It took time for their influence to make itself felt but they were the instigators of the comprehensive system of conservation that eventually prevailed.

Intermission

Thus at the end of the 1930s new forces were gathering to drive forward the cause of conservation. But, despite the successive legislative activities of the previous fifty years, the means of implementation were seriously defective and little used. Apart from ancient monuments, there was still no national system of identifying and listing buildings of architectural or historic importance. Most remarkable was the fact that, despite all the references to buildings of historic or architectural interest in successive Planning Acts from 1909 to 1932, the responsible Ministry in Whitehall had not issued a word of policy guidance or practical advice on the subject.

Despite the growth of scholarly interest in the architectural heritage, that interest hardly extended beyond the best work of the best architects. The SPAB remained set in the distant past and the Georgian Group was distinctly elitist in its approach. Public concern had begun to focus on the picturesque aspects of the English village but was less attuned to the urban scene, and little interest was as yet shown in the domestic architecture of the Georgian, let alone Victorian periods.

There is no knowing how rapidly these concerns might have progressed, or what the government response might have been, had not the Second World War intervened and wrought far more widespread destruction. Five years of war caused a halt both to the destructive course of commercial redevelopment and to the incipient efforts towards conservation. But it also gave a new impetus to public interest in the architectural heritage and concern for its conservation. In 1940 Walter Godfrey and John Summerson, on their own initiative, established the National Buildings Record, which rapidly built up a collection of photographs and scaled drawings of historic buildings and which later formed the basis of the National Monuments Record.

The second part of this study traces the continuing policy history from 1944 onwards. There were to be more dramatic changes in the next thirty years than in the previous sixty. As with many other reforms that began to take shape during the war years, a new era in conservation began with the Town and Country Planning Act 1944.

NOTES

1. Ellis, C.W. (1928) *England and the Octopus* (privately printed 1928; new edition 1975). Glasgow: Blackie.
2. Sykes, C. (1985) *Private Palaces: Life in the Great London Houses.* London: Chatto and Windus.
3. Barker, F. and Jackson, P. (1990) *The History of London in Maps.* London: Barrie & Jenkins, pp. 134–135.
4. Byron, R. (1937) How we celebrated the Coronation. *Architectural Review*, later published as a pamphlet.
5. Hobhouse, H. (1971) *Lost London.* London: Macmillan (revised edition 1976).
6. Summerson, J. (1945) *Georgian London.* London: Pleiades Books (revised edition 1978, London: Penguin).
7. Sproule, A. (1982) *Lost Houses of Britain*, Newton Abbot: David and Charles.

PART 2
1940–1975

6

Prelude

The first part of this study traced the policy and legislative history of urban conservation from its early origins through to the start of the Second World War. Part 2 continues the story from the Town and Country Planning Acts of 1944 and 1947 up to European Architectural Heritage Year in 1975, after which, according to Alan Dobby writing in 1978, 'conservation in Britain was past its peak'. In fact this proved to be a pessimistic view, as will be shown in Part 4, which carries the story forward to 1996.

We have seen how antiquarian interest in ancient monuments became apparent in the seventeenth century, leading to the establishment of the Society of Antiquaries in 1717. In the eighteenth century cultivated taste was more concerned with the rediscovery of classical architecture than with prehistoric or mediaeval buildings in Britain. Interest in mediaeval architecture was stimulated by the activities of the Victorian church restorers and led to the setting up of the Society for the Protection of Ancient Buildings in 1878.

Despite the growing interest in the nineteenth century in the country's archaeological and architectural heritage, the government was extremely reluctant to take any action to protect it. At every stage of the legislative history, it was not the government of the day but a succession of individuals who took the initiative and endeavoured to put in place a statutory system of conservation such as already existed in most European countries. First came Sir John Lubbock, whose untiring efforts led eventually to the Ancient Monuments Protection Act 1882. Then Mr Morrell, a Private Member, who succeeded in inserting into the Housing, Town Planning Etc. Act 1909 a reference to 'The preservation of objects of historical interest or natural beauty' as one of the items to be included in planning schemes made under the new Act. It is not clear how the Housing Etc. Act 1923 came to include the strikingly comprehensive provisions in Section 23, but it seems most likely (and there are some scraps of correspondence in the PRO which support this inference) that it was the result of lobbying by some of the leading professional planners of the day, including Sir Raymond Unwin. But the Ministry of Health still baulked at issuing any official guidance on this aspect of the planning process, just as it had done following the 1909 Act. Finally, while the Town and Country Planning Act 1932 was working its way through the House of Commons, Viscount Cranborne MP attempted to incorporate provisions that would have required the setting up of an advisory committee to compile maps or lists of historic buildings worthy of preservation. Fifteen years later Lord Cranborne, by then in the House of Lords as the Marquess of Salisbury, had another opportunity to press his proposals, again on his own initiative and contrary to the government's original intention – as we shall see in the next chapter.

7

The New System

Town and Country Planning Act 1944[1]

The Town and Country Planning Act 1943 made the planning provisions of the 1932 Act mandatory over the whole country (previously the obligation to prepare Town Planning Schemes applied only to local authorities of over 20,000 population). Thus the ground was laid for the comprehensive post-war planning system introduced by the Town and Country Planning Acts of 1944 and 1947. Historic buildings and conservation were, of course, only a small component of the post-war planning legislation but those two Acts began the process of establishing the system of conservation that exists today. It was, however, a faltering start and it took another twenty years to complete the structure.

There was a brief trailer on 9 March 1944 when the Minister of Town and Country Planning (Mr W.S. Morrison, who later became Speaker of the House of Commons) was asked in a Parliamentary Question by Mr Colegate what powers he had to preserve buildings of national importance. The Minister replied that, apart from his powers under Section 17 of the 1932 Act to approve building preservation orders made by local authorities, he had no such powers. He went on to say, however, that 'The provision of further powers for this purpose is under consideration in connection with further legislation.'

Two months later Mr Morrison introduced the Town and Country Planning Bill. The Bill began its Parliamentary passage somewhat inauspiciously. Mr Morrison started his Second Reading speech by saying: 'In opening this discussion I should like to try to put this Bill in its proper place', at which Dr Haden Guest intervened loudly 'In the wastepaper basket'. Dr Guest, however, did not object to any particular part of the Bill but complained that it was not comprehensive enough – 'one of a series of very bad improvisations in planning'.

The Bill was concerned chiefly with post-war reconstruction, the powers of local authorities to acquire land for that purpose, and the basis of compensation. As originally introduced the Bill contained no provisions relating to buildings of architectural or historic interest. But one Member, Mr Keeling (MP for Twickenham), made a strong speech on the subject. He urged that enthusiasm for reconstruction should not lead to the demolition of historic buildings, and pleaded in particular for the protection of Georgian buildings, to which many local authorities, he claimed, paid less regard than they did to 'older buildings which have inferior claims to preservation'. He proposed that local authorities should be required (not merely empowered) to prepare lists of historic buildings, that the Minister should have power to add to such lists (though not to prepare them himself) and that, pending the preparation of these lists, all buildings earlier than 1850 should be treated as listed (though this should not prevent buildings of later date being included in the statutory lists).

Holland House, Kensington (1605) after bomb damage in 1943. The remains were incorporated in a youth hostel.

He also proposed that local authorities should be required to advertise any proposal to demolish a listed building. He mentioned incidentally that some local authorities had already prepared such lists, including Dagenham, Ilford, Leytonstone, Brighton, Chichester, Cheltenham and his own borough of Twickenham – an interesting band of pioneers.

At Committee stage Mr Keeling moved a series of amendments intended to afford protection to such buildings in the course of redevelopment. For the government, the Parliamentary Secretary, Mr Strauss, expressed sympathy with the objective but asked Mr Keeling to withdraw his amendments on the assurance that the government would consult bodies concerned with conservation and bring forward amendments at Report stage.

This resulted in a new clause which for the first time empowered the Minister, 'with a view to the guidance of local authorities in the performance of their functions under the Town and Country Planning Act 1932, to compile lists of buildings of special architectural or historic interest, or to approve with or without modifications, such lists compiled by other persons.' Before compiling such lists the

Minister was required to consult 'such persons or bodies of persons as appear to him appropriate as having special knowledge of or interest in buildings of architectural and historic interest.'

Thus, once again, a crucial component of the British conservation system was introduced not on the government's own initiative but in response to the initiative of a Private member (backed up by the SPAB, the Georgian Group and other bodies).

The new clause was generally welcomed but several amendments to it were strongly urged: that owners would be notified of the intention to list and before the lists were finalized; that listings should extend to areas adjacent to listed buildings; that the final lists should be published (not merely deposited); and, most importantly, that the Minister should have not merely the power but the *duty* to compile such lists – i.e. that 'may' should become 'shall'. At this stage the government resisted all these proposals.

Later in Report Stage, again in response to representations, the government introduced another new clause requiring owners of listed buildings to notify the local authority of any intention to alter or demolish a listed building. This again prompted demands that owners should receive two months' prior notice of listing. The Solicitor General strongly resisted this proposal on the grounds that it would 'give the owner full time to do all the harm he likes and in that way undo all the good which our proposals bring about.' The amendment was then negatived without further debate.

When the Bill reached the House of Lords some of these matters were raised again in Committee – in particular the question of giving owners prior notification of listing. Several noble lords supported the Earl of Warwick's amendment to this effect. The Lord Chancellor admitted candidly that 'As a matter of fact, I am not quite persuaded what is fair here.' He pointed out that the Minister had to consult expert opinion before listing, but 'On the other hand, I speak frankly and say that I do see the pinch of the argument of the other side.' He asked for more time to consider the point, and the amendment was withdrawn.

At Report Stage in the Lords the government introduced a new amendment requiring that owners should be notified when, but not before, their buildings had been listed – which, surprisingly, the Bill had not previously required. The Earl of Warwick and his supporters tried again to provide for prior notification, but the Lord Chancellor argued against it, and with more conviction this time. In addition to the key point that to give the owner two months' prior notification of listing might prove to be an invitation to demolish, he pointed out that an owner who found that his property had been listed could ask the Minister to use his power to amend the list by deleting that building. He concluded that prior notification 'would make the whole thing unworkable.' The critics were not satisfied but the government drew the debate to a close and the proposal was negatived.

Lord Esher sought to provide that the lists should be published, but he withdrew his amendment on being given the somewhat evasive assurance that they would be published 'from time to time as far as is reasonable and convenient.' In fact the lists were never published in the sense of being obtainable from HMSO, although lists for each local authority area were put on deposit for public inspection.

One other amendment of interest was made when the Lord Chancellor acknowledged that the 1932 Act 'contained a limitation which, when you look at it, is quite absurd.' He was referring to the fact that building preservation orders (b.p.o) could be made to prevent the demolition of a building but not its alteration. As he said 'You might knock it about as much as you please.' The Bill remedied that defect. But the government also accepted an amendment that required notification of intent to alter a listed building only where the proposed works would 'seriously affect the character of the building'. This notification was to enable the local authority to consider making a building preservation order: there was nothing comparable to the system of listed building consent which was not introduced until 1967.

The 1944 Act received Royal Assent on 17 November 1944. It represented some significant advances in so far as the case for listing historic buildings was now acknowledged, but neither local authorities nor the Minister were required to compile such lists. Those who proposed to demolish or significantly alter a listed building had to notify the local authority of their intention but did not have to seek consent: the local authority's sanction was to serve a building preservation order, but the Minister had no default powers. In short, the 1944 Act did not advance the cause of conservation very far but it helped to pave the way for the Act of 1947 and, in effect, invited those who were concerned for conservation to press for more effective measures.

Town and Country Planning Act 1947[2]

The Town and Country Planning Act 1947 was one of the great legislative achievements of the post-war Labour government. It was also an outstanding intellectual and administrative achievement on the part of those who conceived it, carried it through its legislative process and forward into implementation. The need for a comprehensive planning system was generally accepted by the Coalition government and formed part of the preparations for post-war reconstruction. Most of the 1947 Act was built on the pre-war system and the 1932 Act, and commanded a large measure of cross-party support. But there was no chance that the proposed system of compensation and betterment would command similar support. That revolutionary concept ensured that the Bill would have a long and stormy passage, and that debates would focus almost entirely on that aspect.

Conservation benefited from the basic principle of the Bill, since it meant that no compensation was payable as a result of a building preservation order, or planning controls restricting development, except in certain very limited circumstances where existing use value or established development rights were affected. Where an owner could show that his property was incapable of reasonably beneficial use due to a planning restriction, he could serve a purchase notice requiring the local authority to acquire the property. This made it possible to pursue more effective conservation policies.

Against the massive structure of the 1947 Act and its drastic proposals, the specific provisions concerning conservation attracted virtually no interest. But there were some useful minor improvements and one key change that marked a major advance.

The Bill was presented on 20 December 1946 and the Second Reading debate lasted two days – 29 and 30 January 1947. Introducing the Bill, the Minister of Town and Country Planning, Mr Lewis Silkin (later Lord Silkin) said 'This Bill has been described as the most important for a century. I should not go as far as that, but I do say that it is the most comprehensive and far-reaching planning measure which has ever been placed before this House.' Mr Silkin's speech lasted over two hours. He did not refer to historic buildings conservation and nor did any of those who spoke in the debate. The Second Reading was approved by a majority of 192 (those were the days when a majority was a majority).

The Bill as introduced had two main Clauses concerning historic buildings. Clause 26, concerning building preservation orders, substantially re-enacted Section 42 of the 1944 Act but with three significant changes. First, the Bill in Clause 98(2), as part of the extensive menu of reserve powers, enabled the Minister to make a bpo himself, or direct a local authority to do so, whereas previously he could deal only with orders initiated by a local authority. The Bill also gave him power to confirm a bpo immediately, effective for a period of two months, whereas previously a building could be demolished while the order was going through the formal procedures. Thirdly, however, the Bill as originally introduced enabled the Minister to prevent *any* alteration to an historic building by means of a b.p.o, whereas in the course of the Bill's passage this power was amended so as to retain the 1944 Act's limitations to works that would 'seriously affect the character of the building'. This was perhaps realistic but it reintroduced an element of uncertainty and a potential loophole.

The major change came in Clause 27 of the Bill which largely re-enacted Section 43 of the 1944 Act enabling, but not requiring, the Minister to compile lists of buildings of special architectural or historic interest. It was not until the Bill reached the House of Lords that the crucial change was made by altering one word: 'may' became 'shall', and Clause 27 (section 30 in the Act as finally approved) then read: 'the Minister SHALL compile lists of such buildings . . .' (emphasis added).

This historic amendment was moved by the Marquess of Salisbury, who had attempted something rather similar when, as a Member of the House of Commons, he moved amendments to the 1932 Bill (see chapter 4). He seems to have remembered this rather late in the day because he had to apologize for tabling his amendment in manuscript and too late for the printed Order Paper. Nevertheless, he deserves full credit for his initiative. In introducing his proposal on 1 July 1947 he said

> It is at least doubtful whether the local planning authority are quite the right people to perform this particular function. No doubt they are admirable from the ordinary planning point of view. They may be expected to be skilful and devoted in their work, but the preservation of buildings of historic and architectural interest needs a specialized knowledge which I should have thought could not be always in the possession of a local planning authority. So far as I see it, the danger is not that they will include too much; the danger is that they will include too little, and leave out of their list buildings which experts would include. It may be that it would be important to preserve those buildings from the point of view of our national heritage.

This was probably the first time that the term 'national heritage' was used in Parliament; certainly I have not found an earlier one. No one objected to Lord Salisbury's proposal and the Lord Chancellor accepted it, if a little grudgingly: 'I, too, think that this Amendment may (*sic*) be useful, and I am happy to accept it.'

The only other significant amendment proposed was to raise again the issue of prior notification that had been pressed vigorously in 1944. This time Lord Hylton proposed not that owners should be notified individually of the intention to list but that a full list should be published in the *London Gazette*, and in local newspapers, of all the buildings proposed to be listed in a specific area. This, it was urged, would enable owners to object either to the inclusion or exclusion of their buildings. The Lord Chancellor resisted the idea on the grounds that it would lead to unnecessary delay. He noted that the Ministry had already begun preparing lists in many parts of the country and said that he had been told that 'somewhere between 100,000 and 200,000' buildings were likely to be listed. Prior notification would slow the whole process down. In the event the amendment was withdrawn. The Earl of Radnor remarked, 'I dare say it will work out all right.'

Once again, the cause of conservation was moved forward, in no way thanks to the government but to a private individual who saw more clearly what needed to be done. Lord Salisbury may have been mistaken in thinking that it would be impossible to include 'too much' in the statutory lists of historic buildings, but by his amendment he ensured that the central government assumed full responsibility for listing whereas all the earlier planning legislation since 1909 had circled around the need for conservation without providing this vital prerequisite of an effective system.

The remarkable thing is that even in 1947 the government was still so cautious, reluctant or half-hearted about taking this vital step. It may be that it would in fact have continued with the enabling power that the 1944 Act provided. But it is at least equally likely that it would have failed to pursue the task with any consistency or conviction, had it not been for that crucial amendment whereby 'may' became 'shall'.

Nevertheless, despite this advance, the 1947 Act was still an inadequate vehicle for effective conservation. While owners had to notify the local planning authority of a proposal to demolish a listed building, the only means of stopping them doing so was to make a building preservation order, which was a complex and protracted (and therefore expensive) process. The simple solution was to require owners to apply for specific consent to demolition or alteration. It was another twenty years before legislation was enacted to this end. Yet again one is left to wonder at the slow reluctant progress that Britain made towards a comprehensive method of conservation properly integrated with the planning system.

NOTES

1. 7 and 8 GEO 6, 1944, c.47.
2. 10 and 11 Geo 6, 1947, c.51.

8

Implementing the New Acts

Post-War Reconstruction

The 1947 Act at last provided central government with not just the power but the duty, which it had shown no sign of wanting, to prepare statutory lists of buildings of special architectural or historic interest. We shall see how it went about that task but, naturally enough, it did not rank very high among the many post-war demands on government resources. Nor did it feature prominently among the responsibilities of the fledgling Ministry of Town and Country Planning, which was mainly concerned with getting the new development plan system established, grant aiding post-war reconstruction, starting the New Towns and setting up the Central Land Board to run the compensation and betterment system.

While conservation had not in the early post-war years yet entered the mainstream of planning practice, it is notable that one of the most distinguished participants in the post-war conservation process had himself expressed some ambivalence about its purpose. John Summerson published in 1948, as part of his collection of essays on architecture *Heavenly Mansions*, a lecture that he had delivered in 1946 as part of a course at Bristol University.[1] In this essay, 'The Past in the Future', he began by observing that while preservation 'had been held a worthy thing by some few people in every generation since Alberti (c. 1450)', and that 'Today, a large part of public opinion endorses its propriety', nevertheless 'The subject is, however, subtle and delicate, susceptible of fatuity, hypocrisy, sentimentality of the ugliest sort and downright obstructionism. In its worst form preservation may be a resentful fumbling, a refusal to understand the living shape of things or to give things shape.' In this passage Summerson is expressing the uneasiness which those who see architecture as a living art feel when confronted with the extremism of the arch-conservationists and the nostalgic tendency of popular taste. One can only guess what his feelings might have been when faced with the conservation triumphalism of recent years.

Summerson's elitist view of the architectural heritage was certainly shared by the Georgian Group and by others of cultivated tastes who were drawn to the cause of conservation both in the 1930s and in the early post-war years. It was reflected in the views of the Advisory Committee on Listing and to some extent insulated them from the current of public opinion that was concerned more with the impact of redevelopment on the urban scene and on local communities than with the protection of individual buildings of 'special architectural or historic interest'.

Despite his somewhat rarefied view of the subject, Summerson's main concern in this seminal lecture of 1946 was to emphasize that 'preservation in general is only valuable when it is co-ordinated and related to a plan of positive development.' He went on to enlarge

Exeter Cathedral: the Choir after Baedeker raid (1942).

on 'the possibilities of systematic preservation as part of urban planning.' Summerson declared 'I think we should agree that where radical replanning is possible general sentiments regarding historic associations should not be allowed to be an obstruction.' In an aside he remarked ironically that 'Almshouses are apt to be extremely obstructive.' This was somewhat reckless of him, since as he observed in the following paragraph 'The preservation movement has at the moment (1947) gathered extraordinary momentum.'

When his lecture was published in 1948, Summerson added some observations on the implications of the new powers and duties contained in the 1947 Act.

That this burden should be shouldered by the State is admirable. But it is attended by certain dangers. We must always remember the mixed and more or less imponderable nature of the values involved . . . preservation by legislation is valid only so long as it retains the constant and earnest sanction of a minority of the electorate, as well the tolerance of the majority . . . As a preserver of buildings the

State should be reluctant and critical . . . A wide margin of old buildings may properly be left to the goodwill of owners and tenants, the public spirit of groups and societies, and the discretion of planning authorities. Within that margin there will be tiresome breakages, but it is the testing ground of good faith and of the lively, insistent interest in architecture without which preservation is sterile.

With the most distinguished and sensitive member of the Advisory Committee expressing such ambivalent, though deeply felt, views on the whole business of State intervention in the preservation of the architectural heritage, it is perhaps not surprising that among professional planners conservation was not a prominent concern in the early post-war years. Alan Dobby[2] quotes from Lewis Keeble's *Principles and Practice of Town and Country Planning*, one of the leading textbooks of the time which was frequently reprinted.[3] He dealt with 'preservation' in less than half a page and advised his readers that 'This is a subject on the edge of Land Planning proper, and is of direct importance to it mainly as far as preservation of buildings of merit requires redevelopment proposals to be modified to secure their continued existence.' So preservation was important to planners only in so far as it got in the way of their plans for redevelopment! Keeble went on to claim that 'the cost of adapting' old buildings 'would be greater than demolishing them and replacing them by new buildings', and he urged 'the importance of not prejudicing the future for the sake of the past.' He concluded that 'it would be folly, except in the most exceptional cases, to allow an existing building awkwardly situated to ruin the satisfactory redevelopment of a whole area.'

Keeble's book was first published in 1948, and today it seems almost incredible that a leading professional planner and teacher could utter sentiments of that kind. The third edition was published in 1964 and it is clear that at that time planners were still thinking in terms of comprehensive redevelopment not conservation. Enormous damage was done to historic towns in the 1960s and plans hatched in those years carried through into the 1970s; but by then the counter-attack had already begun.

The slow progress of interest in conservation is reflected in successive editions of John Summerson's classic *Georgian London*.[4] In the Preface to the first edition in 1945 he wrote that research on the building history of London since the seventeenth century 'needs doing now, before the age of reconstruction blots out all that vast quantity of minor evidence which, battered and often derelict, cannot be expected to last long.' In the Preface to the second edition in 1962 (fifteen years after the 1947 Act) he commented despondently 'While the final disintegration of Georgian London proceeds apace, the definitive study of its fabric moves all too slowly.' But by the time of the third edition in 1978 he was able to write that, despite the 'lamentations' in the two previous Prefaces, 'the remains of Georgian London are still substantial and, moreover, less directly threatened than at any time since 1945.' He attributed this partly to the fact that the 1947 Act listing process was proving to be 'increasingly effective', and partly to the introduction of conservation areas under the Civic Amenities Act 1967. But, he added, more fundamental than either of these Acts were 'the changed public attitude towards urban environments, the retreat from wholesale clearance and re-planning, and the recognition that Georgian streets and squares can contribute meaning and retain usefulness in the modern city.' He concluded that 'Inevitably, and not unreasonably, the building stock of Georgian London will continue to be reduced but not, we may hope, uncritically or without appreciation of the values which it inherits and creates.'

It is striking that John Summerson should have felt that such slow progress had been made during the first fifteen years after the 1947 Act, and that it was not until two post-war decades had passed that he was able to feel that the tide had begun to turn. But his testimony is very pertinent since he was himself closely involved in the process of 'listing' that the 1947 Act had at last made a statutory duty on the Minister of Town and Country Planning.

Listing Begins

A certain amount of mystery surrounds the start of listing. As already noted, during the debate on the 1947 Act in the House of Lords, the Lord Chancellor mentioned casually that the Ministry had already begun work on listing and that he had been told that it was estimated that between 100,000 and 200,000 buildings were likely to be statutorily listed.

The records of the Ministry of Town and Country Planning at the PRO on this aspect of its work are numerous but sparse.[5] Over a hundred Departmental files on this subject are held on deposit, and their titles appear to cover most subjects, including draft memoranda to local authorities on the new statutory provisions, briefs for debates, correspondence with the Georgian Group and other interested bodies, the question of listing nineteenth- and twentieth-century buildings, theatres, 'group' value, and other topics that are still current today. But the contents of most of the files is disappointingly thin: often only a page or two of tattered carbon copies and with no continuity. One fat file turned out to contain little but letters from members of the Advisory Committee apologizing for inability to attend the next meeting.[6] One suspects that many of the key papers remained in the personal archives of those involved in the work, including some of the notable 'Whitehall irregulars' who were recruited to serve during the war and in the early post-war years. These included Sir Philip Magnus (later biographer of Gladstone) who was the Assistant Secretary in charge, and Sir (as he later became) Anthony Wagner, who did most of the administrative work while at the same time fulfilling his duties as Richmond Herald in the College of Heralds: he later became Garter King of Arms and head of that strange institution.

Advisory Committee on Listing

The 1944 Act required that the Minister, before compiling, approving or amending the statutory lists of buildings of special architectural or historic interest, should consult expert opinion. For this purpose an Advisory Committee was set up in 1945, under the Chairmanship of Sir Eric Maclagan KCVO, FSA, to advise the Minister on the general administration of the relevant provisions of the Act, with particular reference to the principles to be followed in listing.

A minute by Anthony Wagner dated 28 October 1946 referred to the Committee's membership as including three civil servants (Sir Alfred Clapham, Professor Holford and Mr Chettle) and two staff from the National Buildings Record (John Summerson and Walter Godfrey). Wagner suggested that some 'outside' members should also be appointed to the Advisory Committee (he instanced the architect Professor Albert Richardson, later President of the Royal Academy), and that such expert members should be paid for their services. This latter suggestion was not adopted: the Great and the Good, or the merely expert in their field, were, until very recent years, expected to give their services *pro bono publico*.

Continuing with his thoughts on the future role and composition of the Advisory Committee, Wagner suggested that it would serve as 'both a buffer and a lightning conductor between us and a series of outside bodies not particularly notable in the ordinary way for the calmness and unanimity of their views.' He forecast that 'as soon as the first list is made public, these bodies and their members will certainly fall upon it and examine it minutely with a view to possible shortcomings.'[7]

The main service rendered by the Maclagan Committee was to recommend a system of classifying listed building into three categories:

GRADE 1: Buildings of such importance that their destruction should in no case be allowed.

GRADE 2: Buildings whose preservation is a matter of national interest and whose destruction or alteration should not be undertaken without compelling reason.

GRADE 3: Buildings of architectural or historic importance which, though they do not rise to the degree properly qualified as 'special' to justify statutory listing, so contribute to the general effect or are otherwise of sufficient merit that a planning authority ought, in the preparation or administration of its plan, to record them as an asset worth trying to keep.

The Committee recommended that buildings in Classes 1 and 2 should be put on the statutory list and those in Class 3 on a supplementary list. These recommendations were approved by the Minister although the definitions of the three grades of buildings were not closely adhered to either in the *Instructions to Investigators* or in the Report on MTCP published in 1951.

In practice, the Advisory Committee seem to have been spared much controversy of the kind that Wagner anticipated. One can attribute this largely to the famously suave and diplomatic style of its Chairman, Sir William Holford, who took over the chair when he ceased to be the Ministry's Chief Planner and returned to private practice. But the Committee itself worked harmoniously and consisted of a group of architects, architectural scholars and historians of exceptional eminence, including John Summerson, the Earl of Euston, Professor Galbraith, Professor Webb, W.A. Eden, Walter Godfrey, Goodhart Rendel, Brendon Jones and Marshall Sissons.

Aside from the oversight of the splendidly distinguished Advisory Committee, the task of listing was carried forward by a remarkable band of fourteen Temporary Inspectors recruited for the purpose in 1946, led by a newly appointed Chief Inspector, W.J. Garton. On first acquaintance Garton gave the impression of being a somewhat erratic and disorganized enthusiast for historic buildings but, in practice, he showed considerable, if unorthodox, managerial qualities in leading his mixed bag of Temporary Inspectors (who were soon retitled 'Investigators'). He and Anthony Wagner together worked out the ground-rules for listing, anticipated many of the problems that would occur, and got the whole listing process in motion remarkably quickly and successfully.

A stray unsigned minute among these papers noted that the reason why pre-war planning schemes did not include any reference to preservation of historic buildings 'was due to the simple fact that nobody bothered about this power in the 1932 Act . . . and the war prevented us working out anything of the kind.' Be that as it may, when the opportunity came again after the war, it was seized on enthusiastically by those who then held the reins, both on the Advisory Committee and in the Ministry.

INSTRUCTIONS TO INVESTIGATORS

By far the most interesting document to emerge from this neglected recess of the PRO's archives has been what is possibly the sole surviving copy of the *Instructions to*

Investigators which were drawn up primarily by Wagner and Garton. I discovered a copy in a plain brown envelope dated March 1946.[8]

This memorandum was treated with great secrecy: copies issued to each Investigator had to be signed for individually and the number of copies in circulation was strictly limited. When I became Secretary of the Advisory Committee in 1959 I was told of the *Instructions*' existence but it was gently explained to me that neither my predecessor nor I was entitled to see a copy since they were confidential to the Investigators. Some years later the Chief Investigator lifted the veil on the *Instructions* just sufficiently to let it be known how the grading system worked. But the document itself was never published. Because of its exceptional interest, and the fact that it reflects nothing but credit on its authors, I reproduce extracts at Appendix B, omitting those sections that dealt with purely procedural points and detailed 'pay and rations' aspects of concern only to the Investigators.

Curiously enough, the *Instructions* began with a solecism by informing Investigators that 'Section 42 of the 1944 Act imposes on the Minister of Town and Country Planning the duty of listing buildings of special architectural or historic interest for the guidance of local authorities in the performance of their functions under the Town and Country Planning Acts of 1932 and 1944.' In fact, as we have seen, the 1944 Act did not impose such a duty on the Minister and nor did the 1947 Act until Lord Salisbury succeeded in changing 'may' to 'shall'. This mistake could easily have been discovered since Section 42 of the 1944 Act was reproduced in Appendix 1 to the *Instructions*. However, the fact that those responsible for the work believed that they were carrying out the Minister's statutory duty no doubt lent a further sense of urgency and dedication to the task.

The *Instructions* dealt with the question of ancient monuments in a manner which was tactful but did nothing to ameliorate the anomaly of their being the responsibility of a separate Ministry (as they had been since 1882). At that time, however, relations between the two sets of officials were reasonably harmonious and both were housed in the same premises in Onslow Gardens (by 1959 the Listed buildings work had been moved to Chester Terrace in Regent's Park, close to Sir William Holford's own house). The *Instructions* recorded that the Ancient Monuments branch of the Ministry of Works and the Historic Buildings section of the Ministry of Town and Country Planning 'are acting in the closest concert and must continue to do so.' The *Instructions* explained the ancient monuments legislation and noted that it was not true, as was apparently generally assumed, that monuments post-1714 could not be scheduled (that limitation applied at that time to the Royal Commission on Historical Monuments). It was also noted that 'Scheduling is not confined, as some suppose, to pre-historic earthworks, which bulk so largely in the lists simply because they happen to be numerous.' The *Instructions*, however, did not attempt to draw any line between what should be treated as ancient monuments and what should be regarded as buildings of special architectural or historic interest. That grey area remained and we shall return to it later (see chapter 14).

There is no need to describe in detail the contents of the *Instructions,* since they can be read in Appendix B. As will be seen, chapter 3 of the *Instructions* contained detailed guidance on 'The Varieties of Special Interest' that the lists might cover, and, while explaining how particular buildings might be of greater architectural than historic interest, declared robustly that 'Whether the interest of a building whose importance derives from its place in the history of architecture should be called architectural or historical or both is a matter of no importance.'

Chapter 3 also dealt in detail with the important topic of 'group interest', where the individual buildings in a group might not be of *special* architectural or historic interest but where the group as a whole did have such interest and where all the buildings in the group should therefore be listed. However, Investigators were warned that

> There is a nice distinction to be observed here: a building must not be listed merely because its neighbours are good and one is afraid that if it were to be demolished it would be replaced by some vulgar monstrosity which would not be tolerable, and least of all in good company. The design of new buildings can and should be controlled under general planning powers and it is not therefore proper to place on the statutory list a building under Section 42 simply to ensure a congruity of neighbourhood which should as well be achieved in a new building. It must be possible to say that the old building, however plain and ordinary in its kind, has some quality in relation to the context which no new building could have.

This kind of aesthetic discrimination was typical of the highly cultivated and enlightened approach that was adopted in the early days of listing. That same sense of discrimination and selectivity led to the introduction of the Supplementary Lists, which were to include those buildings or groups of buildings which did not qualify for the Statutory List but which should be brought to the attention of the local planning authority as warranting 'special consideration and protection' in the normal exercise of planning control. Investigators were therefore asked 'to exclude from their draft Statutory Lists but to note on Supplementary Lists buildings which have in their view cumulative group or character value, but which have not that degree of intrinsic architectural or historic interest which would naturally be called special interest.'

This question of 'group value' continued to concern the Advisory Committee over the following years, and led them eventually to recommend that Statutory Listing should be extended to groups whose importance became increasingly apparent as the process of wholesale redevelopment gathered pace in the 1950s. It was found that inclusion in the Supplementary Lists was of little value in curbing the 'modernizing' ambitions of many local authorities or the predatory instincts of commercial developers. Eventually, when the Statutory Lists came to be revised in the 1980s, many buildings were transferred from the Supplementary to the Statutory lists, although a residue remained to be included in non-statutory 'local lists' compiled by local authorities.

Chapter 4 of the *Instructions* dealt with 'The Field Techniques of Listing' and it was here that the principles of 'grading' were introduced. These represented some modifications to the principles set out by the Maclagan Committee and had, presumably, been endorsed by the Holford Advisory Committee. The *Instructions* explained that

> In Grade I should be placed buildings of such importance that their destruction should in no case be allowed; in Grade II buildings whose preservation should be regarded as a matter of national interest so that though it may be that now and then the preservation of a Grade II building will have to give way before some other yet more important consideration of planning or the like, yet the Ministry will, in each case, take such steps as are in its power to avoid the necessity of this and where no conflict of national interest can be shown will take such positive steps as are open to it to secure the building's preservation.

It is interesting to see that, at the time when the *Instructions* were prepared, it was envisaged that the Ministry itself would actively intervene to secure the preservation of Grade II buildings. In practice the number of occasions when the Ministry felt obliged to initiate a building preservation order was very limited.

The *Instructions* explained thirdly that

> In Grade III will be placed (1) buildings of architectural or historical interest which do not, however, rise to the degree properly described as special, (2) buildings which so contribute to a general effect that the Planning Authority ought in the preparation and administration of its plan to regard this effect as an asset worth trying to keep.

Finally, the *Instructions* envisaged a fourth category, being those buildings which the Investigator would not consider worthy of inclusion even in the Supplementary List but for which he 'may occasionally believe or know' a case for listing may be made 'by someone else'. Investigators were advised that 'In such a case he could note it but grade it IV, simply in order that the Ministry, if criticised for its exclusion, may have the material for an answer to the critic.' In practice, it does not appear that this ghostly Grade IV was ever evoked.

On the question of listing buildings of more recent date than those covered by the Royal Commission on Historical Monuments (which at that time stopped at 1714), the *Instructions* noted that there was no statutory time limit, but cautioned that 'we must of course be increasingly selective as the present day is approached.' They continued as follows:

> It may be said very roughly that down to about 1725 buildings should be listed which survive in anything like original condition. Between that date and 1800 the greater number of buildings should probably be listed though selection will still be necessary. Between 1800 and 1850 listing should be confined to buildings of definite quality and character. From 1850 down to 1914 only outstanding works should be included and since 1914 none unless the case seems very strong and it appears possible that the building may not be brought to light by central research. It is, however, desirable that the selection of buildings for the last 150 years should comprise without fail the principal works of the principal architects and to some extent it may be possible to ensure this by central research.

In practice, the original lists contained very few buildings later than about 1850, partly because the study of Victorian architecture was at that time very limited (apart from the Gothic revival), and even less appreciated by the general public. The reference to 'central research' in the *Instructions* was inspired by hope rather than actuality, since there was no such resource. The only central expertise available resided with the Chief Investigator and his Deputy, Anthony Dale (who later succeeded Garton as Chief Investigator), and their time was fully occupied with vetting the draft lists prepared by the Investigators in the field. In 1959 the Advisory Committee set up a Sub-Committee chaired by John Summerson and including Nikolaus Pevsner and Mark Girouard (later additions to the Advisory Committee) to review the basis for listing nineteenth-century buildings. Work on listing buildings of the twentieth century, and particularly post-World War II, was not seriously pursued until the 1980s and is still in progress.

As will be seen from the *Instructions* reproduced in Appendix B, a variety of other practical and succinct advice was given to the Investigators for the execution of their novel responsibilities. It is a matter for speculation whether the publication of the *Instructions* would have helped or hindered the work in progress. No doubt, had it been decided to publish them, a good deal of censorship or redrafting would have been thought necessary, which could well have robbed them of much of their candour and directness. What is certain, is that there could be no question of keeping them under wraps in the conditions of even modified 'open government' that prevail today.

Interim Lists

Inevitably, questions soon began to be asked about how long it would take to complete the statutory lists for the whole country. Ministers were pressed to offer forecasts of completion, and no doubt the Treasury wanted to know how long the tiny team of Investigators (mostly on temporary contracts and at salaries well under £1000 a year) would remain a charge on public funds.

In June 1950, the Chief Investigator reported that about 700 local authority districts (out of 1470) had been surveyed to date. At this rate it would take another five years or so to complete the job. But he gallantly offered the suggestion that 'interim lists' could be compiled internally from existing reference books and other sources – at the rate of about one week per district. It appears that he and his Deputy would undertake the work themselves, calling on temporary help from individual Investigators as available. They clearly were inspired by a sense of urgency. As Garton remarked 'The threat of demolition to buildings does not lie with the famous ones but with the lesser known ones.' His proposal was adopted although, in putting it to Ministers, his superior officer, Mr Orpwood, warned that 'We shall have to take some chances and we will, I expect, make a few blunders. This will lead to extra work later on to put things right, which cannot be helped.' One can but applaud this exercise in improvization, which carries something of the imprint of war-time Whitehall when the art of cutting corners was regularly employed.

By February 1951, 828 local authority areas had been surveyed, 348 statutory and 277 draft lists had been issued, 203 draft lists were completed but not issued, and 649 areas remained to be surveyed.

The preparation of the Interim Lists slowed progress on the Statutory lists and to some extent relieved the sense of urgency. Eight years later, the Ministry of Housing and Local Government recorded in its Annual Report for 1959 (Cmnd 1027) that listing had been completed in 931 local authority areas and 'partly completed' in 275. By that date 73,310 buildings had been statutorily listed. At some stage a refinement was introduced whereby a very small proportion of outstanding Grade II buildings were further distinguished by having an asterisk or 'star' added to their grading. I have been unable to ascertain when this device was first introduced; it has not been precisely defined and seems to have no special significance.

Full coverage of the whole country was not completed until 1966. Nevertheless, this second Doomsday Book exercise was a remarkable achievement by the tiny band of Investigators and has been insufficiently celebrated. It has always struck me that Nikolaus Pevsner was extremely grudging in his acknowledgement of the help that he must have received from the Statutory lists in compiling his own *Buildings of England*: in general he acknowledges the lists only when referring to a building not noted in any other source.

NOTES

1. Summerson, J. (1948) *Heavenly Mansions and Other Essays on Architecture*, London: Cresset.
2. Dobby, A. (1978) *Conservation and Planning*. London: Hutchinson, p. 98.
3. Keeble, L. (1948) Principles and practice of town and country planning. *Estates Gazette*, pp. 315–316.
4. Summerson, J. (1945) *Georgian London*. London, Pleiades Books (second edition 1962, third edition 1978., London: Penguin).
5. PRO HLG 103 1–127.
6. PRO HLG 103/1100.
7. PRO HLG 103/13.
8. PRO HLG 104/422.

9

Conservation for Some

The National Trust

Since its foundation in 1895 (see chapter 3) the National Trust had been gradually building up its land holdings and had acquired a small number of historic buildings. Until the 1930s it remained, as its founders had intended, concerned chiefly for the protection of the rural landscape. In the decade before World War II, however, it became drawn into the business of preserving historic country houses. The Trust has always had on its Council members who owned such properties and were naturally aware of the problems that owners faced in maintaining them. It was felt that the properties were best preserved if their owners were able to remain in occupation, especially if they were descendants of the original owners. But this became increasingly difficult as hereditary estates were hit by rising estate (or 'death') duty. In 1904 estate duty stood at 8 per cent; by 1919 it was 40 per cent and in 1930, 50 per cent. At the same time the revenues from estates had declined due to the agricultural depression. There was a real danger that the hereditary estates that maintained historic country houses would disintegrate.

It was this situation that led to the setting up of the National Trust's 'country houses scheme'.[1] Under this scheme an owner could transfer ownership of the property to the National Trust, permanently and inalienably, while remaining in occupation, subject to allowing a measure of public access (in general, for a minimum of thirty days a year), and thus be relieved of liability to estate duty on the property. The owner was also required to provide an endowment sufficient to maintain the property. This was originally calculated on the basis of the sum required to bring the property into good order rather than in anticipation of what might be required in the future. The result has been that some of the properties acquired in the early years have proved to be under-endowed, while more recently the size of endowment required has acted as a deterrent to further transfers.

The legislation required to establish this scheme was introduced by way of Private Bills, leading to the National Trust Acts of 1937 and 1939 (the latter dealt with the case of entailed estates – i.e. those where the owner was prevented from transferring the property away from the family, as was the case at Knole).

By 1944 the Trust had acquired about twenty houses under the country house scheme, and those were not all of outstanding architectural quality. Twenty-six others had been offered to the Trust but, for various reasons, only three of these were eventually acquired. Although the country house scheme later attracted more owners, as taxes rose even higher, at the end of the war it did not seem to offer a general solution to the country house problem. Some owners were unable or unwilling to comply with the terms of the scheme, some properties were not of the quality required by the Trust, and some owners, who were unable to sell or maintain their property, either neglected or abandoned it. This was the situation when the Gowers Committee was set up in 1948.

The Gowers Committee

While the listing process was getting under way, the Treasury set up in December 1948 an *ad hoc* Committee[2]

> To consider and report what general arrangements might be made by the Government for the preservation, maintenance and use of houses of outstanding historic or architectural interest which might otherwise not be preserved, including, where desirable, the preservation of a house and its contents as an entity.

The origins of this Committee, and the motive of the Treasury in setting it up, are obscure. It might seem from its terms of reference that the Treasury had not heard of the provisions in the 1947 Act – or possibly they were concerned that the new process of statutory listing, coupled with the long established arrangements for ancient monuments, might lead to ever-increasing demands on public funds. In fact, however, the initiative seems to have sprung from a genuine concern about the condition of many historic country houses and the difficulties faced by their owners in maintaining them. It is a credit to the post-war Labour government, struggling with all the problems of reconstruction, that they should have devoted any attention to a problem that affected such an exclusive class of property owners. But they did.

The Gowers Committee was a rather odd group of seven representatives of the Great and the Good, none of whom had previously been conspicuously involved in this field (although it included one architect, W.H. Ansell, and the young Professor A.F. Blunt who was already Curator of the Queen's Pictures but not yet known for his espionage interests). The Committee was serviced by the Treasury and attracted a vast quantity of written and oral evidence from every organization concerned and a host of individuals. The annexes to its report provide by far the clearest and most detailed account of the arrangements then in force for the protection of historic buildings.

Although the committee's main concern was with the fiscal and financial aspects of maintaining this part of the national heritage (it was careful to confine its attention to historic houses and chiefly to those in rural areas), it took a broad overview of the subject and its recommendations included some that went far beyond the specific questions of taxation.

In particular, the committee was clearly puzzled by, and became impatient with, the split of responsibilities between the Ministry of Works and the Ministry of Town and Country Planning, and with the two parallel and overlapping statutory systems of preservation. In a splendidly outspoken passage of their report (para 67(ii)) they declared

> It is absurd that the preservation of historic buildings should depend on two largely independent codes, overlapping at some points but differing in the Departments responsible for them, the manner of their administration, the powers they confer, and the types of building to which they apply. It is even more absurd that the question of which code applies should depend, as it sometimes does, on which Department finds itself there first.

This situation led the committee to recommend the setting up of a new separate body, to be called the Historic Buildings Council (one for England and Wales and one for Scotland) that would be 'entrusted with duties both general and specific for furthering the preservation of houses of outstanding historic or architectural interest.' These councils were to be appointed by the Chancellor of the Exchequer and should 'become the central authorities for advising government departments, the planning authorities and owners and others on all matters relating to historic buildings and their contents.' The committee also recommended that steps should be taken 'to get rid of the confusion caused by the existence of two separate sets of statutory provisions for the preservation of historic houses', and, in particular, that the new councils should take over responsibility for compiling the lists of buildings of special historic or architectural interest under the 1947 Planning Act.

Those recommendations that impinged on questions of Departmental responsibilities no doubt became the subject of fierce inter-Departmental in-fighting. The sounds of this battle are now lost in the depths of time and there are no echoes of it to be heard in the PRO. In the event, despite the Gowers Committee's withering criticisms, the muddled responsibilities were not resolved and were to remain in that state until the Ministry of Public Buildings and Works was absorbed, along with its responsibilities for ancient monuments, into the Department of the Environment in 1970 (and were later transferred to the Department of National Heritage, as part of a botched compromise, in 1992).

The Historic Buildings Council for England

The recommendations of the Gowers Committee were largely adopted by the government and incorporated in the Historic Buildings and Ancient Monuments Act 1953.[3] This enabled the setting up of the Historic Buildings Council (HBC) for England, whose main task was to advise on grants payable under the Act to owners of buildings of 'outstanding' historic or architectural interest. Its terms of reference also required it to advise the Minister of Works on the preparation of lists of 'outstanding' buildings. This has all the appearance of an attempt by the Ministry of Works to take over at least part of the Ministry of Housing and Local Government's responsibility for preparing lists under the 1947 Act. But in fact the HBC did not fulfil that part of its terms of reference and the Advisory Committee on Listing continued to oversee the work of listing for MHLG. Sir William Holford and John Summerson were members of both the HBC and the Advisory Committee, and their suave personalities no doubt helped to avoid conflict. But the 1953 Act and the creation of the HBC, responsible to the Minister of Works, perpetuated the on-going confusion of Ministerial responsibility for the heritage which had been so strongly criticized by the Gowers Committee and indeed became part of that heritage.

The Historic Buildings Council comprised a suitable team of grandees and scholars. The Chairman was the Rt. Hon. Sir Alan Lascalles, the Deputy Chairman, the Rt. Hon. J. Chuter Ede MP, and the other members (besides Holford and Summerson) were the Earl of Euston, the Countess of Radnor, Miss D.M. Elliott JP, Christopher Hussey (Editor of *Country Life*), Sir James Mann (archaeologist) and W.M.F. Vane MP (surveyor).

The HBC, following its statutory duty, produced its first annual report in 1953, after only two months' existence and continued to publish them until it was absorbed by English Heritage in 1982. At first they were confined to a record of grants made and brief notes on the buildings concerned. In the first few years the statistics were:

	Number of Applications for Grant	*Grants Made*	*Total Value*
1954	342	86	£268,000
1955	476	112	£371,000
1959	346	90	£1,739,000

In 1956 a system of 'town grants' was introduced, whereby the HBC made grants to certain local authorities on condition that the authority also made a financial contribution. The first recipients were Bath, Brighton and Hove, and King's Lynn. The amounts were very modest. For example, King's Lynn was to receive £1,500 a year from HBC and the council agreed to contribute £750 a year (the product of a penny rate).

Gradually the annual reports came to include more general comments on conservation. In the early years there was always a section on 'Shortage of Domestic Staff', since the Council were very pre-occupied with, and sympathetic to, the problems of the country-house owner. By 1957, however, the Council began to show a wider concern and were perhaps increasingly conscious that their activities were benefiting only a narrow band of private property owners. They deplored the fact that their grant-giving role was restricted to 'outstanding' buildings:

> To a large part of the public, including visitors from abroad, one of the most treasured things in English architecture is the vernacular of the 16C to 19C, the period when most of our villages and smaller towns took on their present shape . . . From the public point of view it is, of course, the widespread existence of the vernacular which makes 'the picture' of England. How much of 'the picture' can reasonably be preserved and how it can be preserved are questions which are now very pressing.

The government responded to this concern not by widening the scope and increasing the budget of the HBC but by giving local authorities similar grant-making powers in the Local Authorities (Historic Buildings) Act 1962, which extended to smaller authorities the powers already available to larger local authorities and simplified the procedures.[4]

In their report for 1962 the HBC drew attention to the corpus of Victorian architecture and gave examples ranging from mansions such as Thoresby and Kelham to Gilbert Scott's Foreign Office and St Pancras Station. They observed that 'It is only in comparatively recent years that such buildings have regained a reputation for architectural merit, or acquired claims to historic interest.' They were therefore conscious that in advising on an application for a grant for a building of this period, while it must no longer be dismissed as worthless, 'the methods and standards which we normally apply in deciding our recommendations may need modification.' The reasons for this were partly the enormous number of nineteenth century buildings still surviving, and partly the very high cost of repairing or replacing the type of ornamental features that were originally relatively cheap to produce. However, they had felt able during the past year to recommend grants to 'several of the better examples of the Victorian mansion' and would continue to do so 'where after careful consideration we feel that an incontrovertible case can be made out.' They expressed relief, however, that they were not responsible for grant-aiding churches, which represented a large proportion of the work of Victorian architects.

1963 marked the HBC's tenth anniversary and they looked back with some satisfaction on what they had been able to achieve over that period, with grants to date totalling £3,826,800. But since 1959 they had been 'rationed' to £400,000 a year and the demand was constantly increasing, with some buildings that had already received grants in earlier years now applying for a second helping. Meanwhile the National Trust had become a major customer and had received grants of £760,000 for 44 properties. Moreover, building costs had risen by at least

25 per cent since 1959. The HBC received an extra £50,000 in the following year but that was quickly overtaken by inflation and stringency in public expenditure.

The HBC continued its Annual Reports in much the same modest format for the next twenty years and they provide a continuous record of the buildings that might have fallen into decay and ruin but for the government's assistance. They also year by year express concern about the huge quantity of historic buildings, towns and villages that could receive no grant aid both because of the limits on the HBC's budget and because the council could not bring themselves to classify them as 'outstanding'. The only solutions the HBC could think of were for legislation to increase the scope of their work and a corresponding increase in their budget – neither of which suggestions commended themselves to the government.

In 1966 the functions of the Minister of Public Building and Works under Part 1 of the 1953 Act (except those relating to ancient monuments) were transferred to the Minister of Housing and Local Government. The HBC moved with them and the Advisory Committee on Listing became a committee of the HBC. In their report for that year the HBC welcomed the move which, they believed, would make it easier 'to evolve a coherent policy for the preservation of historic buildings in the widest sense.'

The 1973 report – the year of the HBC's twentieth anniversary – showed no change in format or contents. It again drew attention to the constraints which the HBC had been deprecating for years – the limited scope of its work and the even greater limitations on its budget. Nevertheless, the cumulative effect of its activities naturally continued to grow. Since 1953 over £9 million in grants had been made to assist preserving 2,392 buildings, and 41 town schemes were in operation. The budget for 1972–73 was over £1 million.

Despite its useful work the HBC does not seem to have been a very influential body. Its membership was always comprised of distinctly superior persons: the Ministry of Works had a penchant for appointing noble, or nearly noble, Chairmen – Sir Alan Lascelles was succeeded first by Lord Hailes and then by Lord Glendoran, and he by Lord Montagu of Beaulieu. Their statutory limitation to buildings of 'outstanding' quality bred an air of exclusiveness with which the Council itself was clearly uncomfortable; and their grants went almost exclusively to the homes of country gentlemen and those grander properties acquired by the National Trust. While the HBC in their annual reports expressed concern for vernacular building and the typical English village or market town, they could not do much about them. To that extent they were some distance removed from the mainstream of growing public concern for conservation. They represented a continuation of the elitist element in the English conservation movement and were unrepresentative of the many other interests involved. For these reasons they were not very effective in conveying those concerns to government.

A Lone Critic

Throughout the long years when the cause of conservation gradually gathered strength, although it struggled to make its voice heard, hardly a voice of dissent was raised. Property owners and developers could readily dismiss conservationist pressure, but conservation as an objective of public policy was generally taken for granted. We have seen (in chapter 2) how in the early days of the SPAB the maverick Robert Kerr kept up a sustained critique of the Society's 'most indiscriminate' philosophy. In 1973 David Eversley (an academic planner, later head of the Greater London Council's Planning Department, who died in 1995) struck a different line of criticism, which is referred to in all the bibliographies on the subject, no doubt because it was almost the sole example available until that time of what might be called 'anti-conservation'.

In *The Planner in Society: the Changing Role of a Profession* Eversley embarked on a sustained invective against the conservationist establishment.[5] He began by observing that 'There is an important and noisy movement (albeit from a minority) for the preservation of almost every scrap of urban equipment that is more than about fifty years old.' He noted that 'the planner' is accused of being responsible for the disappearance of the greater part of our historical urban fabric and cultural heritage, and that the Greater London Council stands accused of 'a systematic attempt to bulldoze away every reminder of the past'. Eversley urged that 'The planner's role here must be to steer a sensible middle course, but the direction must be towards improvement, not preservation.'

Eversley went on to attack 'the extraordinary tastes of that small group of people who constitute, for instance, the Historic Buildings Council' – whom he then listed in detail as though their aristocratic titles and academic qualifications were alone sufficient to condemn them. These are the people 'who dictate what is good and beautiful according to aesthetic standards known only to themselves.' Eversley continued:

> A great deal of what is now supposed to be our priceless heritage was regarded as extravagant Victorian rubbish forty years ago, and may well be again so regarded a generation hence . . . The planner's task is not to set up his own aesthetic yardsticks, half-way between the preciousness of the knightly dictators and the total lack of all taste by the great majority of the population, but to point out the costs of conservation and the benefits of destruction – as they affect the whole community.

Warming to his theme, Eversley went on to suggest a connection between leading conservationists and 'the military–industrial complex' and 'foreign capitalists'. At this point, if not before, the reader may find the argument eluding him. It is a pity that the one uninhibited attempt to question the conservationist philosophy became bogged down in its own rhetoric. The subject would benefit from open debate, but not in those terms.

NOTES

1. Jenkins, J. and P. James, (1994) *From Acorn to Oak Tree: The Growth of the National Trust 1895–1994*. London: Macmillan, pp. 74–103.
2. H.M. Treasury (1950) *Report of the Committee on Houses of Outstanding Historic or Architectural Interest* (Chairman, Sir Ernest Gowers). London: HMSO.
3. 1 and 2 Eliz 2 c.49.
4. 3, 10 and 11 Eliz 2 c.36.
5. Eversley, D. (1973) *The Planner in Society: The Changing Role of the Profession*. London: Faber.

10

A Policy Vacuum

Casework before Policy

Despite the prodigious effort put into compiling the statutory lists, there was a remarkable absence of policy on the wider aspects of conservation. To those who for over a decade had been urging the need for the compilation of such lists as the basis for an effective system of conservation, the lists seemed to be almost an end in themselves. The Ministry of Town and Country Planning followed the tradition set by the old Ministry of Health in issuing no guidance on the subject. Between 1945 and 1951 the new Ministry issued nearly a hundred circulars on the implementation of the 1944 and 1947 Planning Acts, but none made any reference to buildings of architectural or historic importance, other than to note in Circular No. 40 (16 April 1948) that the statutory lists were now being prepared and were likely to be completed for urban areas by early in 1949 and for rural areas in 1951 (an estimate that proved to be out by about fifteen years).

No advice was given to local planning authorities about the purpose of listing or what account should be taken of buildings of architectural or historic interest in the course of preparing the new development plans or in the exercise of development control. Presumably it was thought that, as the Minister had to be notified of any proposal to alter or demolish a listed building, adequate control could be exercised centrally by the Ministry, without the need to promulgate any general guidelines to local planning authorities.

That the Ministry produced no policy advice on the subject of conservation is perhaps accounted for by the fact that those who had campaigned for effective powers of building preservation had not themselves gone far in articulating the reasons for such action by the State. Presumably it was assumed to be self-evident that a nation should safeguard its heritage of fine buildings, as it did its collections of paintings and sculpture. Public opinion was also increasingly conscious of the need to preserve historic towns and traditional landscapes. The Planning Acts provided the mechanism for doing so. Evidently it was not felt necessary to express those objectives in policy terms.

Professional planners also neglected this subject. Early pioneers in planning were much more interested in urban design, garden suburbs and new towns than in preservation. The Planning Acts from 1909 to 1932 focused attention on the planning of areas for new development. In his popular Pelican book of 1940, *Town Planning*, Thomas Sharp wrote at length about 'the English tradition in towns', about both planned and vernacular townscapes, and about how those qualities might be recaptured in new development.[1] But the only reference to preservation in his Index was to 'Preservation – rural'.

It remains surprising that the Ministry had nothing to say on the subject. Nor did the Advisory Committee feel moved to express views on the matter, beyond formulating and refining the criteria for listing. The Ministry issued copious advice on other topics, including a detailed botanical list of trees suitable for 'The Planting in Roads and Streets in Urban and Suburban Areas' (Circular No. 24, 14 May 1946). There was also a charming circular (No. 72, 14 April 1949) on the 'Colour of Telephone Kiosks', which recorded the unanimous agreement reached between the Ministry, the Post Office, the Royal Fine Art Commission and the CPRE that telephone kiosks should be in uniform Post Office Red, except where 'considerations of exceptional beauty or architectural interest may justify special treatment', in which cases the Post Office would supply kiosks in 'battleship grey' with glazing bars in Post Office Red.

The Advisory Committee represented the Ministry's main source of expert advice on listed buildings, and the Royal Commission on Historical Monuments and the leading interest groups (i.e. the SPAB, the Georgian Group and the British Council on Archaeology) were routinely consulted on proposals to alter or demolish them. But for the first decade or so after the passing of the 1947 Act their attention seemed to be focused almost entirely on the individual buildings concerned. Neither the Ministry nor the Advisory Committee seemed to give much attention to the place of listed buildings as part of the built environment, or to their integration into the practice of town planning, despite the fact that William Holford was one of the leading town planners of the time.

In the absence of any widespread interest in the broader aspects of conservation among town planners, it is perhaps not unreasonable that the Ministry chose not to pontificate on the subject but to concentrate on the basic task of completing the statutory lists. As those lists continued to spread across the country, however, the system naturally generated a growing volume of case-work demanding the attention of both the Ministry and the Advisory Committee.

Progress with Listing

Progress with listing and the ensuing casework can be traced through the Report of the Ministry of Town and Country Planning for 1943–51 and subsequently in the annual reports of the Ministry of Housing and Local Government, which in 1952 superseded the short-lived Ministry of Local Government and Planning that had briefly replaced MTCP in 1951.[2]

The 1943–51 Report (which served as the obituary for MTCP) included a section on 'Preservation of Buildings' which described the listing process and for the first time explained publicly the system of classification, but in terms rather different from those used in the *Instructions to Investigators*. The differences in definition may not have been intended to be significant but they appear to imply a rather less rigorous policy:

Class 1 buildings of such importance that only the greatest necessity would justify their removal.

Class 2 buildings which for one or more of the reasons mentioned above i.e. architectural, historical, social etc. 'have a good claim to survival'.

On the other hand Class 3 buildings (those on the supplementary lists) were described with some enthusiasm as 'sturdily built cottages of all periods, unpretentious squares of late eighteenth-century or Regency dwellings, rows of dignified houses without remarkable features' – which suggests that questions were beginning to be asked about the exclusive character of the statutory lists: indeed, buildings of these kinds would almost certainly by now be included in the statutory lists. Local planning authorities were 'asked, whenever possible, to arrange to preserve them in their schemes of redevelopment.'

The progress of listing from 1946 onwards can be seen from the statistics given in the annual reports.

(Cumulative figures)	*1946–51*	*1964*
Listing complete (LA areas)	321	1114
Provisional or partly completed	278	208
Buildings covered	12,496	85,753

In the 1951 Report it was said that it was hoped to complete at least interim statutory lists for all 1,470 local authority areas in England by the end of that year, but in fact it took another fifteen years to complete the job – by which time re-surveys had begun and thousands more buildings were being added to the lists.

Building Preservation Orders

The 1951 Report noted that only thirty-five building preservation orders had been made by local authorities and confirmed by the Minister under the 1932 Act powers by the end of 1950, and seventeen under the 1947 Act. As the number of listed buildings increased, so inevitably did the number of notifications to the Minister of proposals for their alteration or demolition:

	1959	*1964*
Alterations	1582	1678
Demolitions	505	639

Such notices, however, rarely led to the making of building preservation orders.

	1959	*1964*
bpos confirmed	23	39
No. of buildings	46	87
bpos not confirmed	N/A	7
No. of buildings	N/A	30

By 1959 a total of 187 bpos had been confirmed, covering 485 buildings, and by 1964 these figures had increased to 344 orders covering 1,233 buildings. Tree preservation orders were used much more widely: by 1959, 1606 such orders were in force. The Ministry claimed, however, that many more listed buildings were protected as the result of refusal of planning permission or by conditions attached to permission – and indeed, that was the intention: bpos were meant to be a measure of last resort where owners were intent on demolition (for which planning permission was not required at that time).

Examples

The annual reports record year by year examples of buildings saved from demolition by the Minister's intervention. The 1951 Report simply noted that these ranged from country houses such as Winslow Hall, Buckinghamshire and 'the unique timber-framed buildings' of the Butterwalk, Dartmouth, to 'less celebrated buildings' such as the Crown Hotel, Yeovil, and the ancient Crown Court at Godalming which was to have made way for a road-widening scheme. Similar examples were given each year. Thus in 1959 Chatham House, an early eighteenth-century house in St James' Square, which had been the home of three Prime Ministers (the elder Pitt, Lord Derby and Gladstone) was saved from the Royal Institute of International Affairs who wished to replace it with 'a more convenient one'. Also in 1959 Compton Verney in Warwickshire, built by Sir John Vanbrugh and with alterations by Robert Adam and a stable block by James Gibb, was made the subject of a bpo because it was feared that sale by auction would lead to its demolition.

The case of the village of Lavenham in Suffolk is interesting. It had a high proportion of listed buildings, mostly fifteenth and sixteenth century and mostly in poor repair (as it happens, the Permanent Secretary, Dame Evelyn Sharp, had a weekend cottage there, which no doubt helped). The Ministry co-operated with West Suffolk CC and Corford RDC in considering 'how the special character of the town can be preserved'. This led to the appointment of a consultant architect to carry out a survey and to the preparation of what later became known as a 'town scheme' for the village.

Also in 1959 (which seems to have been a good year for conservation) the Minister confirmed a bpo on forty-three houses in Doughty Street, St Pancras, and in doing so noted 'the importance attached to retaining buildings which have value as part of a group rather than as separate entities.' Thus the Ministry endorsed the emphasis that the Advisory Committee had been urging on 'group value'. Contrary to this, however, in 1962 the Minister declined to approve bpos made by the LCC covering 135 houses in Wimpole Street, Devonshire Place and Harley Street. The annual report records that 'He accepted that parts of this area formed a pleasant architectural unity of simple dignified eighteenth century character but he did not think that this was sufficient to warrant indefinite preservation on the scale proposed as the buildings were no longer convenient for living and working.' This was a case where 'group value' was perhaps difficult to demonstrate but which would have been an obvious candidate for designation as a conservation area: but that mechanism was not introduced until 1967.

'Group value', however, was recognized when the Minister confirmed a bpo on the surviving original buildings in Stratford Place (off Oxford Street and barely a stone's throw from Wimpole Street), 'which is a fine example of late eighteenth century townscape.' The Minister decided that, although the original scheme had suffered unfortunate mutilation, the buildings had strong claims to preservation both individually and because together 'they constitute the substantial remaining part of a lay-out of high quality.'

The 1963 Report recorded the preservation of Bedford Square, 'perhaps the most handsome and certainly the most complete of London squares'. But it also noted that the Minister had decided not to confirm a preservation order on 8–16 Montpelier Row, Lewisham, which was described as 'a late Georgian terrace conspicuous for its position on the edge of Blackheath'. It was explained that 'Despite much local support for the order, the Minister decided not to confirm it. The Inspector who held an enquiry reported that the buildings were not particularly distinguished for their period and that they were becoming outworn and outmoded.' (One wonders what would happen to an Inspector who today described a terrace of late Georgian houses as 'outmoded'!) The report continued

> The Minister in accepting his Inspector's findings said that he did not underrate the contribution the buildings made to the character of the surroundings of Blackheath but that, at a time when decent modern accommodation was badly needed, it made sense to preserve property which was outworn, inconvenient and unduly costly to maintain only where the aesthetic arguments were stronger than in this case.

It is not known what the Advisory Committee or the Georgian Group thought of the

Stratford Place Nos 4–7, London, by R. Edwin (1773): subject of a building preservation order 1959.

Minister's decision, but no doubt they would have fought harder for Bedford Square than for Montpelier Row, Lewisham.

The aura of elitism, not to say snobbery, hangs about these two cases. By contrast, 'a complex of modest seventeenth and eighteenth century houses' at the junction of the High Street and Mill Lane at Benson in Oxfordshire were saved from their owners' wish to replace them by a new block of flats – 'These buildings have some intrinsic interest but their main claim to preservation arises from their prominence near the centre of Benson and the contribution they make to the village scene.' So a sentimental feeling for the English village carried more weight than local concern for a prominent terrace of Georgian houses in Lewisham.

In nearly all the cases cited, the initiative in making the necessary bpos was taken by the local authority (the LCC was especially active) rather than by the Minister who was responsible for approving all such orders but also had power to initiate orders himself. At this time, there was no effective alternative to a bpo: it was not until the 1967 Act that 'listed building consent' was introduced, by which specific permission had to be sought for the alteration or demolition of a listed building, and control of demolition was not extended to unlisted buildings in conservation areas until 1972.

Victorian Architecture

The Ministry's annual report for 1960 recorded that the Advisory Committee had been considering the question of listing buildings of the nineteenth and early twentieth centuries.[3] The architecture of the Victorian period had attracted increasing scholarly interest over the past decade or so, notably in the works of Alfred Russell Hitchcock, Nikolaus Pevsner and Goodhart Rendel. John Betjeman had stimulated interest beyond the academic level. But public opinion was as yet not much concerned for Victorian architecture and much was unceremoniously demolished as commercial redevelopment accelerated in the late 1950s.

The Advisory Committee, after reviewing the existing criteria for listing buildings post-1850, concluded that 'instead of only outstanding works of the period 1850–1914 being listed, buildings of the whole of the nineteenth century and up to 1914 will be listed on the same basis . . .' i.e.

> Between 1800 and 1914 listing should be confined to buildings of definite quality and character, and the selection should include the principal works of the principal architects. Certain factors should also be kept constantly under review; these include special value within certain types, whether for architectural reasons (including planning) or as illustration of social and economic history – e.g. industrial buildings, railway stations, schools, hospitals, theatres, town halls, markets, exchanges, 'pubs' and cemeteries; and value for innovation or virtuosity in the technological field e.g. cast iron, prefabrication, early use of concrete.

The Advisory Committee reported that they had tested the effect of adopting these principles in selected areas, and were satisfied that 'there was no reason to expect an overburdening of the lists.' As there was no early prospect of Investigators revisiting areas already surveyed so as to up-date them in the light of these new considerations, the Advisory Committee themselves drew up a list of the 'principal architects' and recommended that the Ministry should compile by central research lists of their surviving works so that a selection could be made for addition to the lists if not already included. The Ministry accepted these proposals and work was put in hand.

Unfortunately, within a year or two of this initiative, two outstanding Victorian buildings became the focus of public controversy and in both cases the government decided against their preservation. These were the Euston Arch and the Coal Exchange.

Two Landmark Cases

These two cases marked a turning point in attitudes towards conservation. In each case enormous opposition was aroused both from the cognoscenti and from the media when it was learned that they were to be demolished. In both cases the government was unmoved by the conservationist outcry and repeated appeals to save the buildings. At the time it seemed that these two decisions confirmed that the interests of conservation would never command sufficient political importance to stand in the way of major public works schemes. The government seemed indifferent to charges of vandalism and philistinism. And yet it is difficult to think of any major building of comparable importance that has been lost since the time of those decisions in 1961–62.

The two buildings in question were both mid-nineteenth century, and it is true that at that time Victorian architecture, while the subject of much scholarly work and championed by the Victorian Society (formed in 1958), had not yet acquired a large popular following. The Ministers concerned were probably right in sensing that the fuss made about their demolition did not reflect any great public concern. But, in the event, it seems to have marked the turning of the tide which then began to flow in the direction of conservation. The two buildings and the story of their demolition can be briefly summarized.

The Coal Exchange

This was a purpose built structure in Lower Thames Street designed to house the London coal market – i.e. a specialized commodity trading market. It consisted of a high-domed

The Coal Exchange, City of London, by J.B. Bunning (1849); demolished for road widening 1962.

trading floor surrounded by several galleries of traders' offices. Apart from the cast-iron dome structure, it was notable for the abundance of cast-iron decoration using mining motifs and large encaustic panels celebrating the industry. The architect was J.B. Bunning and it was opened by the Prince Consort in 1849. By the 1950s, despite being obsolete, largely unoccupied and somewhat neglected, it was by no means derelict. Professor Russell Hitchcock called it 'the prime city monument of the early Victorian period'. The City of London had long had a scheme for widening Lower Thames Street, and it was claimed that either the Coal Exchange or the Customs House of 1813–17 (which faced it across the road and was still in use as the headquarters building of HM Customs and Excise) would have to go to make way for it. In 1958 the Coal Exchange was listed Grade II (the Custom House by David Laing was already listed Grade I). The Victorian Society persuaded the City Corporation to defer demolition for two years while possible ways of saving the Coal Exchange were worked out, including one by the architect Lord Mottistone which involved running the footpath in arcades for a short stretch of the road at a cost of £125,000. In 1962 the Minister of Housing accepted that this scheme was practical but concluded that 'it would involve drastic alteration to the exterior of the building and would substantially reduce the accommodation provided', and he decided not to intervene. The City Corporation allowed demolition but agreed to meet half the cost of a somewhat desperate last-ditch proposal by the Victorian Society to ship the dome to Australia where it would be re-erected as part of the National Gallery of Victoria at Melbourne. But the Society could not raise the other half of the cost and the Coal Exchange came down in November 1962. For over ten years the site remained vacant while other land was assembled for the road widening, which was eventually completed. Hermione Hobhouse called it 'one of the great conservationist horror stories'. The other two cases that she coupled with the Coal Exchange were the Adelphi (demolished in 1936 to make way for an office block) and the Euston Arch.

The Euston Arch

Euston Station was London's first railway terminus and the first such terminus in any capital city in Europe. The Arch, a huge neoclassical stone structure, was designed by Philip Hardwick (1792–1870) as the entrance to the station and was opened in 1858. The British Transport Commission (predecessor of British Rail) applied to the LCC in 1960 for planning permission to redevelop the station, including both the Arch and the main building by Hardwick's son; both were listed Grade II. The LCC accepted the demolition of the main building (which included the grandiose Great Hall in which Victorian directors of the Great Western Railway held their meetings) but suggested that the Arch should be re-erected nearby. BTC refused to meet the cost of £180,000. A deputation from the Royal Academy, the SPAB, the Georgian Group and the Victorian Society met the Prime Minister, Harold Macmillan, but the government declined to save the Arch and it was demolished at the end of 1962. The demolition contractor, Frank Venturi, numbered and conserved the masonry blocks at his own expense and they ended up in the garden of his home in the country. In 1994 the BBC television programme *One Foot in the Past* rediscovered those remains but also found that about 4000 tons of the Arch had been bought and used to fill a large hole in the bed of the River Lee, from which a huge segment of a Doric column was raised with its fine fluting in perfect condition. The *Architectural Review* (which had led the campaign for the Arch's preservation) in its Editorial in April 1962 wrote of the 'long drawn out history of bureaucratic dilatoriness and evasion', and commented that the Arch's destruction

The Euston Arch, London, by Philip Hardwick (1858): demolished 1962.

> is wanton and unnecessary – connived at by the British Transport Commission, its guardians, by the LCC and the Government, who are jointly responsible . . . one of the outstanding architectural creations of the early 19C, and the most important – and visually satisfying – monument to the Railway Age which Britain pioneered.

No subsequent politician has shown Harold Macmillan's *sang froid* when confronted by the fury of the conservationist lobby. But at the time when the Coal Exchange and the Euston Arch were lost there was no provision for listed building consent; the only remedy available to prevent demolition was a building preservation order, and neither the LCC nor the government was prepared to pay for that. Moreover in both cases it was claimed that demolition was necessary in order to facilitate major public works. In neither case was that strictly unavoidable, but in the early 1960s the post-war enthusiasm for redevelopment had not yet evaporated. The loss of these two buildings, however, and the politicians' acquiescence in their loss, probably fuelled the growing demand for more effective powers of conservation.

'Outrage'

In 1955 there had appeared a book that came to have a profound effect on the attitude of professional planners, and many others, towards post-war planning and redevelopment, and led to a total change of approach, although it did not prevent the demolition of the Coal Exchange and the Euston Arch. This was Ian Nairn's *Outrage*, which originated as a special issue of the *Architectural Review* and was published as a book later the same year.[4] Nairn's primary target was the poor quality and incompetent planning of new development that he felt was destroying both town and country: 'This death by slow decay we have called "subtopia".' It was a term that caught on instantaneously and has entered the language. In a key passage he wrote

> Places are different: Subtopia is the annihilation of the difference by attempting to make one type of scenery standard for town, suburb, countryside and wild. So what has to be done is to maintain and intensify the differences between places. This is the basic principle of visual planning.

Nairn was less than impressed by the notion of 'conservation' that had recently begun to take hold:

> How about historic buildings? Are they to be allowed to look old, or are they given a 'quaint' facelift under the excuse of preservation? – and are the best 18C buildings preserved? or only those which reflect a travel-poster Merrie England of beams and tracery?

Ian Nairn's vituperative style and Gordon Cullen's vigorous illustrations rammed his points home. They followed up their campaign with a continuing series of articles in the *Architects' Journal* that demonstrated both what was wrong and incompetent in current town planning and development, and how far better results could be achieved. Their message got through to the popular press and eventually to government policy and Departmental publications.

A New Impetus

In their report for 1966 the Advisory Committee recorded that they had been 'particularly concerned about the threat to the character of historic towns arising from the scale and pace of urban redevelopment', and that the Ministry was preparing a new publication on 'the problem of preserving the quality of the environment of ancient towns and historic areas, while providing for the changes necessary to meet modern needs.' This publication was to be based on studies of selected towns which would 'illustrate the qualities and types of areas which merit protection and methods of reconciling preservation with change.' This initiative originated in the last year of the Conservative government when Sir Keith Joseph was Minister. It was to be carried forward by the incoming Labour government which resulted in a marked new impetus to conservation policy.

NOTES

1. Sharp, T. (1940) *Town Planning*. London: Penguin (revised editions 1942 and 1945).
2. Cmnd. 8204.
3. Cmnd. 1435.
4. Nairn, I. (1955) *Outrage*. London: Architectural Press.

11

A New Policy Approach

Urban Renewal

In the 1960s the Ministry of Housing and Local Government (MHLG) began to think about the relationship between conservation and planning. It can hardly be said to have done so before 1960. The work on listing was carried on very conscientiously, but it was in something of a backwater so far as the Ministry's main interests were concerned. Immediately after the war the main effort went into reconstruction, starting up the first generation New Towns, and getting the new planning system into operation. In the 1950s the spotlight was on the Conservative government's commitment to build 300,000 houses a year. The planning side became pre-occupied with processing the first round of development plans.

Then in 1961 the Joint Urban Planning Group (JUPG) was set up, bringing together a small staff group from MHLG and the Ministry of Transport (a marriage that was consummated ten years later when the Department of the Environment (DOE) was established – although that ended in divorce in 1978).

The JUPG set about producing a series of Planning Bulletins which represented the Ministry's first serious work on planning policy and technique since the days of Holford and the Ministry of Town and Country Planning in the 1940s. Eight Planning Bulletins were produced between 1962 and 1967. The first in the series was *Town Centres: Approach to Renewal*, which had some sensible things to say about planning and conservation.[1] It was at this time that commercial developers had been let loose on town centre redevelopment after the decline of public-sector-led post-war reconstruction. The early results of this speculative redevelopment were beginning to cause alarm among both conservationists and town planners interested in urban design (in those days many more planners than now had first qualified as architects and some Planning Schools had a tradition of teaching urban design).

Although Planning Bulletin No. 1 was concerned with a wide range of issues, including the new problems of traffic in towns, it reflected the growing concern about the impact of redevelopment on historic town centres. It emphasized the importance of the traditional character, qualities and features that had survived and the historic buildings that had been preserved, and urged that 'Redevelopment should not ignore these qualities; they strengthen the sense of continuity between the past and the present without which a town becomes anonymous and dull.' Two short sections are worth quoting, since they summarize the approach that a few individuals, notably John Summerson, and to some extent the Historic Buildings Council, had been advocating and which sought to reconcile the interests of conservation with the necessary process of planned renewal and redevelopment:

> *Character*: Every town has a character of its own and underlying all other objectives will be the wish to retain and enhance its individuality. Indeed, in some towns the main objective will be to preserve the town centre as far as is humanly possible in its present form . . . A town's character is an elusive quality and it requires sympathy and resourcefulness to ensure that it is not lost in the process of renewal.
>
> *Conservation* is an essential part of the renewal process. The need to retain the good qualities and best features of the town centre has already been emphasised. But this cannot be taken for granted. The features and areas worth preserving have to be identified and active measures taken to ensure that their value is appreciated and where possible enhanced in the process of renewal.

These thoughts may seem commonplace now, but they needed to be given expression at that time and they are certainly still relevant today when conservation has perhaps come to predominate over the interests of renewal.

Planning Bulletin No. 4 *Town Centres: Current Practice* repeated much of what was said in Bulletin No. 1 and attempted to illustrate how those objectives were being achieved in practice.[2] It proved very difficult to find successful examples of creative conservation going beyond the preservation of individual buildings, since there was still more interest in redevelopment than in conservation. The bulletin chose to take an optimistic line on what might be achieved:

> Town Centres today are undergoing more rapid and more sweeping changes than at any time in the past. As redevelopment gathers pace a new idiom of town centre design can be seen evolving – something entirely modern, distinctive and enjoyable.

This was more an expression of hope than of achievement, apart perhaps from some of the New Towns, where entirely new town centres were being designed. There were few examples of skilful renewal combined with conservation, and far too many examples of grossly insensitive commercial redevelopment. The government wanted to encourage the concept of 'partnership' (now an overworked cliché but a new idea at that time) between local authorities and private-sector developers. Some dubious deals were struck, and those developers who were most active in soliciting inexperienced local authorities were not noted for the quality of their design or building. Some towns still bear the scars of those 'partnerships'; a good many listed buildings were lost in the process and a great many more which would have been listed by the standards that prevail today.

These two Planning Bulletins were focused on urban renewal rather than on conservation as such, but they showed evidence of new thinking in the Ministry. This was followed up over the next few years when the new Labour government took office in October 1964. But the ground was prepared during the latter years of the Conservative administration and while Duncan Sandys and Sir Keith Joseph were in charge of MHLG. The result was a fair degree of bi-partisan support for the new policy approach.

NOTES

1. Ministry of Housing and Local Government (1962) *Town Centres: Approach to Renewal*. London: HMSO.
2. Ministry of Housing and Local Government (1963) *Town Centres: Current Practice*. London: HMSO.

12

Ministers Make Policy

The 1960s saw the emergence for the first time of conservation as a policy issue of some political importance and the development of more coherent policies. The credit for this goes to three politicians – Duncan Sandys, Richard Crossman and Lord Kennet (who also used his family name Wayland Young – he was the son of Sir Hilton Young who introduced the 1932 Planning Act – and wrote under the name of Wayland Kennet).

In 1954 John Harvey, the author of *The Conservation of Buildings* declared in a speech to the Ancient Monuments Society that 'there is no real or consistent policy' on conservation and that this reflected 'the absence of any philosophy of conservation.'[1] He continued 'The country faces a challenge, nothing less than the loss of the most tangible part of its traditional heritage . . . action must be immediate. Tomorrow will be too late.'

As we have seen, until 1962 successive governments had never issued any useful policy advice on the subject and, in so far as any policies evolved, they did so through decisions on individual cases, and the results were far from consistent.

Duncan Sandys

Duncan Sandys was Minister of Housing and Local Government in the Conservative administration from 1954 to 1957. His main preoccupation as Departmental Minister was with housing policy. He initiated the first post-war slum clearance drive, which grew rapidly in scale and momentum, and the reform of the Rent Acts. He rightly saw rent control as a main cause of the deterioration of the stock of private rented housing; and his solution was to decontrol private rents (with some exceptions). But he also took a close personal, if somewhat erratic, interest in historic buildings and in urban design. His most notable involvement was with the controversial redevelopment proposals for the bomb-damaged area around St Paul's Cathedral, and he achieved fame in the Department by calling for a saw with which he removed the top storeys of one of the tower blocks in the model of Sir William Holford's plan for the development of Paternoster Square to the north-west of the cathedral (Paternoster Square is now itself the subject of major redevelopment proposals).

In 1957, while still Minister, he was instrumental in setting up the Civic Trust, of which he became President.[2] The aim of the Trust was to stimulate interest in the protection and improvement of historic town centres by establishing local Trusts and by promoting local schemes of restoration. The Permanent Secretary, Dame Evelyn Sharp, was appalled at the idea of her Minister setting up what she saw as a new pressure group that was bound to be a nuisance to the Department. But Duncan Sandys was a stubborn man and he went ahead with it. He did not, however, do anything much to develop the Ministry's own approach to conservation. He played a more active part when he left office following the Conservative party's defeat in the 1964 election.

Richard Crossman

Following the 1964 election, Richard Crossman became Minister of Housing and Local Government. It was not what he wanted or expected. His main interests during some thirty years on the back-benches had been foreign affairs and as front-bench Opposition spokesman on pensions and social security. But he had longed for office and he set about his first Ministerial job like a child in a toy shop – or, as some of his officials (including Dame Evelyn Sharp) felt, like a bull in a china shop. He gives a day-by-day account of his Ministerial experiences in *The Diaries of a Cabinet Minister, Volume One 1964-66*.[3] As it happens, I was his Principal Private Secretary for most of that period.

The Diaries record Crossman's difficulties in coming to grips with the world of government and Whitehall, and especially with his first Department. His first priority was to repeal Duncan Sandys' rent legislation and to make a start on the Labour government's manifesto commitment to increase the rate of council house building. But he soon developed an inexhaustible interest in every other aspect of the Department's activities. He showed a considerable capacity for work and immense intellectual energy, coupled with total naivety about the workings of the Whitehall machine. But in little more than two years he accomplished a great deal. Historic buildings were only a tiny part of the Ministry's work but it did not take long for him to discover it and he brought to it a genuine interest in the subject (though not much knowledge of it). We can trace his activities in this field through *The Diaries*, which provide a unique opportunity to see policy making through the eyes of a Departmental Minister – though hardly a typical one. Janet Morgan's admirable edition of *The Diaries*, with its excellent Index, enables us to do this.

His first mention of historic buildings comes in the entry for 3 March 1965, after six months in the Ministry. He had been having lunch with the Professor of Naval History at Greenwich, for whom he had lectured for years. Professor Bullock wanted to talk to him about the preservation 'not of a single "listed" house but of a group of buildings on Blackheath.' Crossman records that only the previous week he had paid a visit to one of the Ministry's London outposts at Caxton House and met the Townscape Group:

> They were working on a new policy document which suggested that preservation of ancient buildings should be concerned not merely with individual listed buildings but also with groups or streets or small areas of old towns.

Hardly an entirely new idea, since the Advisory Committee and the HBC had already drawn attention to the question of 'group value'. But it was new to Crossman and he took it up with his customary enthusiasm. 'I spent some time trying to persuade Dame Evelyn and J.D. Jones (the Deputy Secretary) that this idea is of great importance and that we should legislate on it.' He did not succeed immediately but he did not let go of the 'idea'.

Two days later on 5 March at his weekly meeting with Dame Evelyn and J.D. Jones, he raised the question again, along with that of gypsies. He 'got nowhere' on either problem. 'I realised that in each of these cases the irritation of the office is due to the mere idea of any intervention by the Minister. They think this is the kind of area which Ministers should leave to them.' Nothing was more calculated to ensure Crossman's continuing interest.

This also enabled him to play a leading part on 26 March in filibustering against a Private Member's Bill which the government wished to defeat but with which some Labour backbenchers were in sympathy. George Wigg suggested that it should be talked out by prolonging the debate on the Consolidated Fund Bill which was scheduled for the same day and on which Private Members can put down motions on any subject they please. Crossman turned up and sat through three hours debate before contributing 'an impromptu seventy-minute' speech on listed buildings and then went home to bed. The debate dragged on and the Private Member's Bill was talked out – for which the government got a very bad press. But it confirmed Crossman's interest in historic buildings.

By 4 April he felt able to record that 'I think I have managed to get the section dealing with "listing" of buildings reorganized. At least I have doubled the clerical staff and reduced the time in which the list can be prepared from ten to three years.' On 6 April he discussed the subject with Lord Holford over lunch before having a haircut (he went to one of the best barbers in London about once a fortnight). Unfortunately he was not much impressed by Holford, who had been Chairman of the Advisory committee since 1953 and was past his best. On 31 July Crossman had lunch with John Betjeman at his house in Wantage and discussed the subject again. Evidently he found this occasion more stimulating. He explained to Betjeman that he had told Holford that

> I wanted to broaden the scope of his committee and try to make it less pedantic and more concerned with real policy planning. Holford is a tired man and a bit of a spent force and he obviously wanted me to kick him out of the chair. I was clever enough to keep him as Chairman while I put the changes through.

Other concerns took priority for the next few months but on the weekend of 8–9 January the following year the Department organized a conference for him at Churchill College, Cambridge which went very well:

> John Betjeman and Tom Driberg (the MP) were there as well as thirty or forty other experts, and Jimmy James (Chief Planner) and J.D. Jones ran the conference brilliantly. At last we are going to be able to work out a good policy for historic towns. Our idea is to select four for pilot experiments, where we not only list individual buildings but plan the preservation of the historic core. One of the four will certainly be Bath. From Cambridge John Delafons motored me back to Prescote (Crossman's family home in Oxfordshire) where we had a weekend of brilliant sunshine mixed with appalling snow showers.

I remember the journey well. I drove Crossman in my ancient Hillman convertible with the hood down. He talked non-stop from Cambridge to Oxford, but not about conservation.

Crossman did not have to deal with many cases involving conservation but he records two in *The Diaries*. On 15 May 1965 during a visit to Newcastle he insisted on seeing Eldon Square 'a beautiful Georgian Square which they are going to destroy for a shopping centre'. He had been advised not to visit the site because the case was *sub judice* in the Department; or, if he must go, he should do so incognito and talk to nobody about it. In the event he walked round it accompanied by a crowd of journalists, talking cheerfully all the time. He was 'passionately opposed' to the demolition, but found that the procedures had already been completed:

> I blew up our regional staff in Newcastle and told them that they were vandals for giving my consent. But I knew that it was already a fait accompli, and when I get back to the Department I shall be forced to draft the directive letter saying that they should have permission.

Crossman was right in his view: Eldon Square was a grievous loss and the eventual redevelopment was horrific. He could have tried to stop it by last-minute reversal of the decision but it would probably have landed in the Courts – where his impromptu on-site comments could have been held to be prejudicial. He was wrong in thinking that he

Eldon Square, Newcastle on Tyne: before redevelopment (1966).

would have to draft the decision letter himself: it was done and signed on his behalf.

The second case showed the Minister intervening in the opposite direction. On 11 May 1966 he went to see a house in Cheyne Walk, Chelsea, which the Greater London Council wanted to preserve as a late example of the art nouveau style of the late nineteenth century.

The owner, as usual, wants it pulled down for redevelopment. When I got there I was surrounded by representatives of the GLC, builders, the contractors and the Ministry. I stalked alone through the house. Outside, it had no conceivable merit; inside, it was a total wreck with the wall-papers rotting off the walls, a staircase utterly worm-eaten, and a little garden house which once may have been nice but in which all the frescos had rotted away. Transformation into flats, which the GLC recommended, would have destroyed the last trace of art-nouveau. I decided to let it be scrapped.

That was one of his last decisions before the General Election of 31 March 1966. Labour returned with a majority of ninety-seven seats; the Conservatives failed to gain a single seat. Crossman was returned again for Coventry East. The Prime Minister, Harold Wilson, made twenty-five changes in the Ministerial team; two Cabinet Ministers and ten non-Cabinet Ministers left the government, but Crossman was retained in his Ministerial seat and professed that this was exactly what he wanted. He got the message from the Prime

Minister on the phone while he was dining at the Garrick. A few minutes later the PM phoned again to say that, in addition to taking over three Ministers from the Ministry of Land and Natural Resources (which was to be re-integrated with MHLG as Crossman wanted) he would have another one. 'I'd forgotten to tell you I want you to have Kennet as well.' 'Kennet!' I said. But what for?' The PM replied 'We want young blood.' Crossman said 'Thank you very much' – 'And that was that.'

So Lord Kennet joined the Ministry of Housing and Local Government. It was his first government post. At first Crossman did not know what to do with him and assumed that he would just speak for the Department in the House of Lords. But Lord Kennet himself has recorded that when the Prime Minister sent for him, he reconciled him to his posting to MHLG (rather than the Foreign Office, as he had hoped) because he would 'be able to look after historic buildings.'[4] Kennet felt pleased about this because ten years earlier he and his wife had written a book, *Old London Churches*, and he had 'retained an interest in the history of architecture and in urban design.' It was in fact an inspired appointment, since Lord Kennet outstayed Crossman at MHLG and made a very substantial contribution to policy on conservation and planning.

Kennet told Crossman what the Prime Minister had said and asked for the historic buildings job.

Crossman let me have it, and this was pleasant of him since he had been doing it himself without assistance from a junior Minister and had, I think, been enjoying it pretty well. It is enjoyable; not many beautiful things come into a Minister's life.

We will resume Lord Kennet's part in the story later, but first we can return to *The Diaries* to follow the final stages of Crossman's interest in conservation and planning during his time at MHLG, in which Lord Kennet was also involved.

On 17 May 1966 Crossman had a meeting with Duncan Sandys who had drawn first place in the ballot for Private Members' Bills. Sandys asked whether the Department had available 'a nice Bill about keeping the countryside clean'. Crossman pressed on him the need for a Bill to strengthen the powers to prevent demolition of listed buildings, and Sandys agreed to adopt this measure. Sandys also agreed with Crossman's concept of townscapes as against the listing of individual buildings, and 'I said that that was marvellous and that we would make a point of getting that into the Bill.'

Crossman asked Sandys to call on him again later that month (20 May):

The Department had done its homework and so had I. I was able to tell him that the Cabinet Committee had given its consent to the main outlines of his Bill. I was also able to tell him that I was still determined to make sure the Bill dealt with townscapes – that is groups of buildings – not merely with individual listed buildings, although the Department had obstinately drafted my policy paper for the H.A.C. excluding this concept. When I said all this, Miss Williams, who is the number two in historic buildings, was enormously excited. She could hardly believe her ears and felt for the first time that something really was going to happen. No doubt this was partly due to the fact that the control of historic buildings was unified under me a few days ago as a result of the Department taking over one section of the Ministry of Works. But largely I think she had the feeling that her Minister was battling and had fought down the resistance of her superiors. It had taken me nearly two years to do it

Crossman quickly got to know Kennet and liked him. He and his wife dined with the Kennets at their house in Bayswater. On 26 May he wrote:

My last meeting that day was a full-scale Ministry conference on historic buildings. Kennet is really splendidly energetic. Though he peeves me a little by his desire to take everything over, I am delighted that he is that sort of person. He is going to run the historic buildings as hard as he possibly can and he is going to be helped by Duncan Sandys' Bill as well as by the decision to get working parties going on my five selected towns. That was a good meeting to have before I went off to Crete.

(where, incidentally, he very nearly bumped into Dame Evelyn Sharp: she noticed him at Knossos and dodged behind a pillar so as not to be seen – or so she told me. I thought it odd that the Minister and his Permanent Secretary had not discovered beforehand that they were both going to the same place for their summer holiday).

During the Summer Recess of 1966 Crossman learned that the Prime Minister was planning a Cabinet reshuffle and that he was to become Lord President of the Council and Leader of the House of Commons. Crossman was pleased with this promotion though it was not a role that he had ever seen himself in. He gave a small farewell party in his office, at which he said of his time in MHLG that 'I had never been happier in all my life but I knew a lot of people would be relieved when I had gone'. He took his family on holiday to Cornwall and on the last day there committed some thoughts to his diary on his time in the Ministry and what he felt he had achieved. In particular he wrote:

> One of the areas in which I took a particular interest was historic buildings. When the 1966 election was won and I got my way, I insisted that there should be a new allocation of functions in the Ministry of Works and I got the whole of the listing of historic buildings and subsidies transferred and centralised in the Department. This kind of work was utterly despised by Dame Evelyn. She regarded it as pure sentimentalism and called it 'preservationism', a word of abuse. She, who counted herself a modern iconoclast, took the extremely – yes, I will say it, – illiterate view that there was a clear-cut conflict between 'modern' planning and 'reactionary' preservation. During my time as Minister, in speech after speech, I tried to break down this false dichotomy and to establish a new and sensible relationship between planning and preservation.
>
> In doing this I found the actual division which deals with historic buildings extremely rigid and difficult. One of them expressed to me a passion for Duncan Sandys and showed himself an extreme political conservative as well. Although I was the first Minister to show a genuine enthusiasm for their subject they continuously resented my interference and tried to defeat me whenever they could. On one occasion, when I tried to get the Karl Marx Library in Clerkenwell listed as an historic building, the vote of the officials present turned me down. Altogether my relations with them were very hostile, in contrast to my relations with the R.I.B.A. and the architects which got more and more friendly. (I got a letter from Lord Esher the other day saying that in their considered view I was the best Minister they had ever had.)

Crossman was succeeded at MHLG by Anthony Greenwood, a weak Minister who some felt took little interest in anything except his personal appearance. He was content to leave most of the work to his junior Ministers and this gave Lord Kennet ample scope to pursue his special interest in historic buildings, as will be seen in the next chapter.

I have quoted extensively from the Crossman *Diaries* partly because he did take a genuine interest in the subject, but also because it shows, what is often doubted, that a determined Minister with ideas of his own and the energy to pursue them can make policy even when his senior officials do not fully share his enthusiasm.

NOTES

1. Harvey, J. (1972) *Conservation in Buildings*. London: John Baker.

2. Esher, L. (1982) *The Continuing Heritage: The Story of the Civic Trust Awards*. London: Freney. This provides a copiously illustrated account of the Civic Trust's admirable work in promoting quality in both conservation and new building.

3. Crossman, R.H.S. (1975) *The Diaries of a Cabinet Minister: Vol.1: 1964–66* (edited by Janet Morgan). London: Hamish Hamilton and Jonathan Cape.

4. Kennet, W. (1972) *Preservation*. London: Temple Smith, p. 49.

13
The Policy Momentum

Civic Amenities Act 1967

Lord Kennet stayed on at MHLG as Parliamentary Secretary from 1966 to 1970 and was able in that time to make a very positive contribution to conservation policy. He gives his own account of those years in his book *Preservation.*[1] It shows that a junior Minister can have a significant influence on policy-making if he takes a keen interest in an aspect of a Department's work that is not of high political importance but which allows him scope to take his own initiatives.

Kennet records how, when he arrived to take up his new post, Crossman was in the process of securing the transfer from the Ministry of Works of responsibility for the Historic Buildings Council and the grants made under the Historic Monuments Act 1953. Crossman had also hoped to take over the work on ancient monuments but the Ministry of Works fought hard to hold onto it, and that transfer was not effected until the Department of the Environment was set up by the Conservative government in 1970 and absorbed the whole of the Ministry of Public Buildings and Works. Kennet remarks that even the transfer of historic buildings to MHLG was not accomplished 'without bloodshed', and that he watched the Whitehall battle 'with awestruck incredulity'.

Once that battle was won, it brought together for the first time Ministerial responsibility for both conservation and planning. It was none too soon. Kennet records how he felt that the climate of public opinion had been changing in recent years after what he called 'an age of blindness':

> There was a general shift of our national consciousness towards the visual, and a greater flow than ever of good and readable scholarship about architecture and the arts. Everywhere local amenity societies sprang to life, and more and more people became conscious that their street, their village, their town, their quarter of a city, was different from others because it had grown differently, and that that was interesting.

Kennet began his new Ministerial job by going to see what was being done about conservation in France and Italy. He learned about the French *zone protégé* which defined a circle one kilometre in diameter around every listed monument and *site classé* (of which almost every town and village had at least one), within which every proposed demolition and new building had to be approved by the central government. This was the system that André Malraux, as Minister of Culture, had introduced in 1962. Kennet did not think that the French system could be readily transferred to England, partly because about four times as many buildings were listed in England as in France (119,000 against 29,000), partly because the methods used to restore the *secteurs sauvegardés* were draconian and often resulted in the displacement of the existing inhabitants, and because the type of work undertaken was 'fabulously expensive'. Nevertheless, he was 'attracted by the idea of increasing central government control at the expense of local government.' He felt that we had much to learn from the French, 'but more from the verve and comprehensiveness of their approach' than from the actual procedures they had adopted. In Italy he found systems similar to the French, but very little progress in restoration except in Sienna. He concluded that in France and Italy the threat to old town centres was one of decay, whereas in England it was development.

Lord Kennet next discovered that the Historic Buildings Council had a budget of only £450,000 a year, and that seven times that amount was spent on military bands. He no doubt seized with enthusiasm on Duncan Sandys' willingness to use his place in the ballot for Private Members' Bills to introduce what became the Civic Amenities Act 1967.[2]

The Bill had its Second Reading on 8 July 1966. Duncan Sandys, introducing it, said that 'In very broad terms the Bill had three purposes: to preserve beauty, to create beauty and to remove ugliness . . . it seeks to protect the character not only of individual buildings of interest but also the area around them.' It was very clear that he had not envisaged conservation areas as extending over whole neighbourhoods or parts of towns, but simply as a means of protecting the setting of individual buildings of note. This indeed had been a concern of both the HBC and the Holford Committee.

It is important to note this relatively limited purpose in view of the way in which the conservation area has come to be used so widely as a means of protecting neighbourhood amenity rather than the immediate setting of important buildings. Section 1 of the Act says (italics added)

> Every local planning authority shall from time to time determine which parts of their area . . . are *areas of special architectural or historic interest* the character or appearance of which it is desirable to preserve or enhance, and shall designate such areas (hereafter referred to as 'Conservation Areas') for the purpose of this section.

It seems rather odd that the draftsman should have used the terminology of listed buildings to define areas which were clearly not intended to be limited to such buildings. In practice this has not been regarded as a restriction on the use of those powers.

The first Member to speak in support of the Bill was Nicholas Ridley, who was later to become responsible for the Act as Secretary of State for the Environment (and who was, contrary to popular misconception, much concerned for the architectural and rural heritage; he was also a very gifted watercolourist). He commented that 'The extension of the principle of listing good buildings to the principle of listing good areas is of immense importance.' He continued 'I welcome very much the provision in the Bill to extend the powers of local authorities for compulsory purchase of buildings that are neglected or threatened.' This sounds uncharacteristic of the Ridley who later showed himself to be so hostile to local government; but, as was usual with him, he showed his consistent political philosophy by urging that once local authorities had acquired and repaired historic buildings they should then sell them.

James MacColl, the Parliamentary Secretary, said 'In general the Government very much welcome the Bill.' But he added 'We must not be preservationists at all costs but must experiment with working old buildings into a modern setting. That is an important part of group preservation.' The Bill went through its subsequent stages without dissent.

'Historic Towns: Preservation and Change'

While the 1967 Act was making its way onto the statute book, the Ministry was working away on developing the new positive approach to planning and conservation which had already made its appearance in Planning Bulletin No. 1 in 1964 (see chapter 10). This work led, in 1967, to quite the best publication that the Ministry (or its successors) ever produced on the subject. This was *Historic Towns: Preservation and Change.*[3] It was printed in unusually large format, well written and very well illustrated (although at that time HMSO had not heard of colour printing).

In his foreword the Minister of Housing and Local Government, Anthony Greenwood, sought to convey the new message about the need to use planning as an instrument of conservation:

> This book . . . is about an aspect of town planning that is now receiving more and more attention: the kind of planning which is needed in order to preserve, in a positive way, the good things our towns already possess. It is concerned not only with single old buildings, but also with the general visual qualities of historic towns . . .
>
> There is of course much in our towns that needs to be changed, and there is no reason why preservation should prevent desirable change . . . The book suggests that the surest way to avoid a conflict between the old and the new is to plan preservation and change together . . .

The message was perhaps a trifle sanguine in suggesting that such conflicts could readily be planned away, and in its confidence that planners knew how to do it. But it did seek to integrate conservation into the planning process as had never been done before, and it is only to be regretted that this objective was not sustained or pursued with much success in later years when conservationists too readily blamed planners for not invariably siding with the cause of preservation rather than seeking jointly to achieve the fusion of 'preservation and change' that the 1967 book sought to demonstrate.

Credit for the seminal book of 1967 must go mainly to the small team in MHLG led by Roy Worskett, an architect–planner who later left the Ministry to become City Architect and Planning Officer of Bath. In 1969 he published his own book *The Character of Towns: An Approach to Conservation*, which developed and illustrated in much more detail the concepts set out in the Ministry's earlier publication.[4]

Worskett placed great emphasis on the need to integrate planning and conservation. In his foreword he wrote 'Above all conservation must be seen within the framework of general planning policies.' In following this in his introductory chapter, he made it clear that he was concerned not only with buildings of special architectural or historic interest but with the character of towns in general: 'This book . . . is about our existing towns, their intrinsic visual qualities, and the differences between them; not just the so-called historic towns but all those with an individual identity.' Thus Worskett was one of the first, possibly the first, to take the idea of 'conservation' beyond the traditional scope of 'preservation' so as to encompass a much wider interest in the historical dimension of towns and cities, and in the sense of time and place as reflected in the urban townscape. This theme was explored and elaborated by those whose interests were more to do with social anthropology than architecture, but also included some such as Kevin Lynch who had an architectural background.[5] Worskett was an innovator in that he was concerned to show that 'conservation policy is not simply a matter of dealing with historic buildings, or areas that contain historic buildings, but is also a part of a creative process that can provide inspiration and discipline for change.' And again 'We must understand physical change and its basic social and economic origins. We must be clear what we want to conserve, and why, and as part of the process of town planning we must develop techniques to cope with such a fundamental conflict.'

Worskett was to find later, during his time at Bath, that it is never easy to resolve this conflict or to satisfy those who are more interested in preventing change than in understanding the process of change. Nevertheless, his work represents the best and sanest approach to conservation and one which still needs to be re-affirmed. The need to reconcile the interests of preservation and the need for new development is too easily reduced to a political cliché. And in truth they cannot be reconciled in the sense of fully achieving both objectives at the same time. But Worskett sought to demonstrate that they were not totally incompatible in the wider context of urban planning.

Four Studies in Conservation

In 1968 MHLG published the reports on four *Studies in Conservation* that were commissioned jointly by the Ministry and the City and County Councils concerned in 1966.[5] The four studies were initiated during the passage of the Civic Amenities Act 1967. The original purpose of the studies was 'to examine how conservation policies might be sensibly implemented in those four historic towns.' The objectives were 'to produce solutions for specific local problems, and to learn lessons of general application to all our historic towns.' By the time the reports were published Anthony Greenwood was Minister but Lord Kennet was still with the Department. In his preface to the studies Mr Greenwood said that the purpose of the studies 'has been to discover how to reconcile our old towns with the twentieth century without actually knocking them down. They are a great cultural asset, and, with the growth of tourism, they are increasingly an economic asset as well.'

The four studies supplemented and illustrated the general approach set out in the Ministry's publication *Historic Towns: Preservation and Change*. The reports of the four studies were:

Bath	by Colin Buchanan and Partners
Chester	by Donald W. Insall and Associates
Chichester	by G.S. Burrows
York	by Viscount Esher

There was originally to have been a fifth study, but the town chosen for that honour, Kings Lynn, would not agree to meet their half-share of the cost of the study, so it did not proceed. Bath and Chester saw it as a good investment, but York was reluctant and was persuaded to join in only when the local Civic Trust offered to pay a quarter of the cost, leaving the city council to meet the last quarter.

The four studies were all produced in a uniform high-quality format, unusual in government publications at that time. They differed considerably in character, reflecting the professional backgrounds of their authors. Buchanan was an engineer/planner who had formerly worked as one of the Ministry's Planning Inspectors and had achieved fame in 1964 as the author of *Traffic in Towns*,[6] which advocated the separation of traffic and people by the building of major urban throughways and the safeguarding of areas where environmental and amenity interests should prevail. Donald Insall was an architect with a high reputation in conservation work. G.S. Burrows was a surveyor/town planner. Lord Esher was an eminent architect and writer, and a founder of the Georgian Group.

The four towns presented somewhat similar problems in terms of the impact of traffic and economic development on the historic fabric of an old city with a wealth of listed buildings.

Three of the studies adopted what would now be regarded as a fairly conventional approach, focusing on the preservation of historic buildings and townscapes, but with less attention to the social and economic life of the community than would now be expected. The exception was Colin Buchanan, who used the study of Bath as an opportunity to apply the concepts of traffic and environmental management set out in *Traffic in Towns*. His report caused something of a furore because it further developed the concept, first set out in his firm's earlier *Bath Planning Study*, of building a major new urban road through the city, partly in a cutting and partly in a tunnel. This was the 'cut route' which would run from New Bond Street in the east to Charles Street in the west, passing underneath the north-south streets such as Queen Street, Barton Street and Princes Street. It would involve substantial demolition and create significant opportunities for redevelopment. The route avoided the most important parts of Bath and relatively few listed buildings would be affected, but these included part of Queen Street which would have to be demolished, as would Nos 16 and 17 Old Bond Street. The consultants suggested that the two Old Bond Street buildings should be rebuilt in replica and that part of Queen Street might be rebuilt behind the existing facades. In terms of its direct impact on listed buildings the effect was therefore quite modest, considering the scale of the proposal. But the whole concept of a tunnel running below Bath, and the scale and character of the approach roads above ground and the wide entry portals, attracted enormous opposition. It was never built and Bath has continued to suffer from the effects of ever-increasing traffic congestion, despite modest traffic-management measures.

Each of the other three towns presented problems of dealing with traffic and the increase in private car ownership. Chichester acquired a by-pass and a large amount of car parking ingeniously inserted on back-land in the city centre (but more recently increased by far more conspicuous surface and multi-storey car parks). York and Chester would rely mainly on traffic management and pedestrianization.

The reports for Chester, Chichester and York all proposed new powers of compulsory conservation and new grants for the purpose. Donald Insell remarked that 'Both government and public are becoming increasingly uneasy because the remarkable planning machine now worked out in Britain still provides so little opportunity for the public initiative needed to protect the past.' He proposed a new system of listing based on a dozen criteria, and to include post-1914 buildings and 'street and pavement surfaces'. There would be two types of list, one a 'planning list' including all buildings of architectural and historic importance, and a 'conservation list' of buildings selected from the first list and which would receive full statutory protection. Lord Esher also found the existing lists (which had been twenty years in the making) unsatisfactory because, while 'invaluable', they were 'designed to convey information and not to imply action.' He proposed a new system of four categories:

A. Ancient monuments and buildings of national importance or major local interest.
B. Buildings of considerable architectural importance but not quite in the first category.
C. Good buildings of major landscape value.
D. Buildings of some landscape value.

Buildings in the first three grades were to be preserved and attract grant, but 'To extend absolute protection to Class D would in our opinion stifle imaginative design and stop in its tracks the process of history which the streets of York so wonderfully illustrate.'

Few of the more general recommendations in the four reports were adopted by the government, and certainly no-one wanted to embark on a totally new system of listing. But the four reports set a new standard of planning for conservation and demonstrated how sensitive detailed analysis of an historic town, with a subtle appreciation of its townscape qualities as well as its individual historic buildings, could lead to creative policies for its preservation and enhancement. The four studies (leaving aside Buchanan's radical solution) were no doubt a major influence on the practice of planning for conservation. They illustrated a positive and imaginative approach that allowed for change and renewal but based on sound technique and respect for local feeling. That approach gradually came to prevail over the next twenty years or so, although it often encountered less discriminating demands for preservation and resistance to change.

Circular 53/67

Once the Civil Amenities Act had received Royal Assent, and the 1967 book was published, and the four town studies had been started, Lord Kennet still retained his keen interest in conservation and was able to keep up the policy momentum in the Ministry.

It was thanks to him that the conservation area concept took hold so quickly and, some would later argue, was applied so indiscriminately. Some of the officials in MHLG were inclined to proceed more cautiously, but Kennet wanted a robust approach: 'I wanted the local authorities to designate many and large areas, which they probably would if they did so before thinking out what had to be done, and then apply increasingly energetic policies in them. If they first discovered how much had to be done and how troublesome it was all going to be, and only then began to designate the areas in the light of that newly acquired knowledge, they would no doubt designate more sparingly. So we advised the local authorities to designate first and think later, and this, broadly speaking, is what they did.'[7]

The Departmental circular on the 1967 Act (MHLG Circular 53/67 issued on 7 August 1967) did not promulgate the new policy on conservation areas in quite the boisterous terms that Lord Kennet employed, but it was certainly less equivocal than most Ministerial policy statements. No doubt Lord Kennet took a close interest in its drafting. It marked a dramatic departure from the mute attitude that previous Departmental circulars on the Planning Acts had adopted towards this subject since 1909. That reticent approach had already been overtaken by Planning Bulletin No.1 and was left further behind in *Historic Towns: Preservation and Change* in 1967.

Departmental circulars are usually distinctly circumspect. Circular 53/67 was a new departure, as the following brief extracts show (the references to 'the Ministers' denote that the new circular emanated from both MHLG and the Welsh Office):

> The Ministers attach particular importance to the designation of conservation areas, which represent a shift of emphasis from negative control to creative planning for preservation.
>
> Preservation should not be thought of as a purely negative process or as an impediment to progress . . . The destruction of listed buildings is very seldom necessary for the sake of improvement; more often it is the result of neglect, or of failure to appreciate good architecture.
>
> Clearly there can be no standard specification for conservation areas. The statutory definition is 'areas of special architectural or historic interest the character or appearance of which it is desirable to preserve or enhance', and these will naturally be of many different kinds. They may be large or small, from whole town centres to squares, terraces and smaller groups of buildings. They will often be centred on listed buildings, but not always; pleasant groups of other buildings, open spaces, trees, a historic street pattern, a village green or features of archaeological interest, may also contribute to the special character of an area. It is the character of

areas rather than individual buildings, that section 1 of the Act seeks to preserve.

Conservation areas will however be numerous. It is for this reason that the Act requires all local planning authorities to designate them. The Ministers hope that they will all take early steps to establish areas.

The Ministers will watch progress with interest, and intend to review the position in about 12 months' time. They would hope that by then every local planning authority will have made an effective start in implementing section 1 of the Act, and that conservation areas will have been established in all the more important historic towns.

Those who wonder when the floodgates of conservation were opened, need look no further than Circular 53/67.

By 1972 Lord Kennet was able to record that some 1,350 conservation areas had been designated by about 130 local authorities. Twenty years later that total had risen to over 6,500 and questions were being asked about the validity of the process. But there is no doubt that it was a political success and attracted very widespread support among those whose interest in conservation had more to do with neighbourhood amenity and private property values than with preservation in the traditional sense.

The Town and Country Planning Act 1968

In the year following the 1967 Act the government introduced the Town and Country Planning Act 1968.[8] The main purpose of this Act was to implement the recommendations of the Planning Advisory Group's report on *The Future of Development Plans* which had been published in 1965.[9] This replaced the 1947 Act style of development plans with the system of county structure plans and more detailed local plans that remains in place today. But Part V of the Act contained new provisions on historic buildings and conservation which came into force on 1 January 1969.

MHLG Circular 61/68 (4 December 1968) explained the provisions of Part V in great detail (thus setting the precedent for later circulars on this subject, which are distinguished by their prolixity). The main feature of Part V was the introduction of the system of listed building consent. Previously the only means of preventing the demolition of a listed building, or those types of alteration that were not covered by normal planning control, was to make a building preservation order. Under the 1968 Act all such works of demolition or alteration affecting a listed building would require specific consent from the local planning authority (or from the Minister if he called in the case for his own decision or if it came to him on appeal following refusal of consent by the planning authority). Unlisted buildings in conservation areas were made subject to a similar system of 'conservation area consent', which controls total or partial demolition but not interior alterations.

Circular 61/68 introduced an entirely new doctrine – the 'presumption in favour of preservation'. This doctrine was expounded in the exaggerated terms that came to typify conservation policy as it moved rapidly away from the balanced approach reflected in *Preservation and Change*:

Generally it should be remembered that the number of buildings of special architectural and historic interest is limited, that the number has already been reduced by demolition, especially since 1945; and that unless this process is, if not completely arrested, at least slowed down, the virtually complete disappearance of all listed buildings can be predicted within a calculable time. Accordingly, the presumption should be in favour of preservation except where a strong case can be made out for a grant of consent after application of the criteria mentioned.

Those 'criteria' were the importance of the building, its historical interest as well as its architectural merit, the condition of the building and the cost of repairing it, and the importance of any alternative use for the site. The last of these criteria introduced a somewhat counter-conservation notion that was not repeated in later circulars on the subject – 'whether the use of the site for some public purpose would make it possible to enhance the environment and especially to enhance other listed buildings in the area; or whether, in a rundown area, a limited redevelopment might bring new life and make other listed buildings more economically viable.'

This pragmatic approach was lost sight of over the coming years but no-one was allowed to forget the 'presumption in favour of preservation'.

The 1968 Act contained a host of minor procedural improvements on which Circular 61/68 commented in detail. But it is not the purpose of this study to chronicle the detailed legal provisions. The significant fact is that policy now favoured the tightening of conservation control to the point of rigidity.

Finally, the circular announced that 'The Ministers expect those authorities who have still not designated all their conservation areas to complete the process as quickly as possible.' The notion that there was a finite number of conservation areas awaiting designation proved misconceived. Twenty-five years later, local planning authorities are still designating new conservation areas and extending old ones.

The Preservation Policy Group

Meanwhile Lord Kennet took the chair of a new Group that was formed to coordinate the four town studies and to consider their implications. It was a powerful group and included Sir Nikolaus Pevsner, Theo Crosby the architect and designer, and Professor Alan Day the economist, together with some of the best practitioners from the local authority world.

Although this policy group had wide terms of reference, including the possible need for new legislation and financial measures, its eventual report published in May 1970 was rather disappointing.[10] It included useful summaries of the four town studies and information about recent developments in conservation policy in Great Britain and in four European countries. The report commented that public attitudes towards conservation in the UK had been changing rapidly, 'and we do not think it would be an exaggeration to say that there has been a revolution over the past five years in the way old buildings are regarded and in the importance now attached by public opinion to preservation and conservation.' It is interesting to note that at this point in time the two words 'preservation' and 'conservation' were both used in this context but the distinction between them was not explained. Although Lord Kennet seems to have preferred 'preservation', and used that as the title of his book in 1972, it seems increasingly to have been accepted that 'conservation' implies a broader approach.

The report had little to recommend in the way of new legislation: 'What are needed are not more powers but the will, the skill, and the money to use the powers that already exist'. The report did, however, propose two legislative changes. First, that a local authority should be able to recover from owners the cost of any emergency repair work that it undertook to prevent further deterioration of a listed building under Section 6 of the Civic Amenities Act 1967. Second, that Section 53 of the Town and Country Planning Act 1968 should be amended so as to exclude from compensation the 'break-up' value of a building which the local authority had compulsorily acquired to ensure its preservation (the 1968 Act excluded development value but had left this loophole).

The report also proposed a new form of grant, comparable to the grants already available for comprehensive redevelopment in areas which it proposed should be called General Conservation Schemes, on the analogy of General Improvement Areas under the Housing Act 1969. As the Working Group observed, it seemed somewhat anomalous that government grants were available for comprehensive redevelopment but not for comprehensive schemes of conservation. For the rest, the report's recommendations were confined to minor administrative improvements, training and good practice in conservation.

If the 1970 report was somewhat disappointing in its limited recommendations, this was evidence of the advances made in conservation policy and in the legislative provisions over the past five years. If Lord Kennet and his hand-picked policy group could not come up with any more radical proposals at this stage, it was unlikely that anyone else could have done so – other than the perennial need for more tax-payer's money.

Lord Kennet was authorized to announce on 20 May 1970 that the government accepted the policy group's report. In particular, 'as a first step', the amount of grant which the Historic Buildings Council could recommend was to be increased from £75,000 a year to £700,000. In addition 'At an early opportunity the Government will introduce legislation to enable the payment of the proposed new grant on general conservation schemes.' It was envisaged that the new grant, together with the HBC's existing scheme, would build up gradually to £1½ million in 1973–74.

Before these proposals could be put into effect, however, the general election of May 1970 intervened. Labour lost the election and Lord Kennet was not to enjoy Ministerial office again. Nevertheless, he could certainly view his work for conservation (or 'preservation' as he preferred) over the past five years with considerable satisfaction. It was a remarkable achievement for a junior Minister, and it can be said that no other single Minister before or since has done as much for conservation as Lord Kennet achieved from 1965 to 1970. Part of his legacy, however, was to be a ball and chain – the 'presumption in favour of preservation'.

NOTES

1. Kennet, W. (1972) *Preservation*. London: Temple Smith.
2. 2 Eliz c.69.
3. Ministry of Housing and Local Government (1967) *Historic Towns: Preservation and Change*. London: HMSO.
4. Worskett, R. (1969) *The Character of Towns: An Approach to Renewal*. London: Architectural Press.
5. HMSO, 1968.
6. Bucahnan, C. (1964) *Traffic in Towns*. London: Penguin.
7. Kennet (1972), *op. cit.*, p. 66.
8. 2 Eliz c. 72.
9. Ministry of Housing and Local Government 1965) *The Future of Development Plans*. London: HMSO.
10. Ministry of Housing and Local Government (1970) *First Report of the Preservation Policy Group*. London: HMSO.

14

Continuity of Policy

Those who may have feared that the new Conservative government elected in May 1970 would mean a lessening of the commitment to conservation that the Labour government had shown, need not have worried. If anything, the new political importance attached to conservation increased over the next four years.

The Department of the Environment

One of the new government's first actions was to produce a White Paper on *The Reorganisation of Central Government.*[1] Among the changes announced was the setting up of the Department of the Environment, which came into being in the autumn of 1970. The foundation for this had been laid during the previous Labour administration, when Anthony Crosland was appointed in October 1969 as Secretary of State for Local Government and Regional Development, with responsibility in Cabinet for MHLG and the Ministry of Transport, and with a remit from the Prime Minister to consider the case for further integration. In 1970 the two Ministries were combined and the Ministry of Public Building and Works also joined the new Department of the Environment, with Mr (now Lord) Peter Walker as Secretary of State.[2]

The new Department was something of a monster, with 52,500 staff. Of these, the ex-MHLG staff numbered only 4,800 but they brought with them by far the most varied range of policy work. The next two years were ones of major organizational change. One of the most complex of these was to combine the former Ministry of Works' responsibilities for ancient monuments and Royal Palaces with MHLG's work on listed buildings and conservation areas. The result was a new combined Directorate of Ancient Monuments and Historic Buildings. Thus for the first time the two separate conservation regimes were brought together in one Department of State. But for a good many years the two constituent parts, each with its distinct history and objectives, continued their separate lives.

New Acts and New Circulars

There now followed a sequence of legislation and policy circulars that added substantially to the conservation system. It is not clear what prompted all this activity but it demonstrated that the new DOE intended to treat conservation as an important part of its work.

The Town and Country Planning Act 1971 consolidated most of the earlier Planning Acts but also incorporated some amendments, including Section 277(8) which provided that 'Where any area is for the time being designated as a conservation area special attention shall be paid to the desirability of preserving or enhancing its character or appearance' in exercising planning control and similar procedures. This apparently anodyne precept seems to have caused little difficulty until a series of legal cases arose, starting in 1988, in which the lawyers had a fine time arguing whether development could be permitted if it neither preserved nor enhanced a conservation area but left it unchanged or unharmed. This weighty issue was eventually resolved on appeal to the House of Lords when Lord Bridge ruled that 'The statutorily desirable object of preserving the character or appearance of an area is achieved either by a positive contribution to preservation or by development which left character or appearance unharmed, that was to say preserved' (*South Lakeland District Council v. Secretary of State for the Environment and Carlisle Diocesan Parsonages Board (1991)*). Be that as it may, Section 277(8) – now Section 72(1) of the Planning (Listed Buildings and Conservation Areas) Act 1990 – provides a comprehensive mantra for the conservation cause and one that is invoked with increasing frequency, although the judgements implied remain essentially subject: the only necessary requirement is that the local planning authority or Inspector deciding the case makes it clear that they have taken account of Section 72(1)/1990.

The Town and Country Planning (Amendment) Act 1972 contained two principal provisions.[3] The first introduced for the first time control over the demolition of unlisted buildings in conservation areas (section 8), and the second provided government grants for 'outstanding' conservation areas (i.e. those identified as such by the Historic Buildings Council).

The Act did not apply demolition control to all conservation areas but only to those where the local planning authority made a direction to that effect. The accompanying circular 86/72 extended the 'presumption in favour of preservation' almost as far as it could reach: 'In Circular 61/68 it was stated that in relation to listed buildings the presumption should be in favour of preservation except where a strong case could be made out for the grant of consent. In the same way, the presumption should normally be in favour of demolition control in a conservation area. The decision whether to allow demolition of a building subject to a direction must depend, however, on the facts of the case and the value of the building to the area.'

A year later in August 1973 a new circular (46/73) was issued under the title *Conservation and Preservation*, although it did not attempt to clarify the distinction between those two terms. Its main purpose was to draw local authorities' attention to a remarkable statement that Lord Sandford, one of the DOE's junior Ministers, had made in a debate in the House of Lords in January on 'The Quality of Life'. Lord Sandford told their Lordships that

> We need full and firm conservation policies. Past and present conservation work has concentrated upon areas of exceptionally important landscapes and historic townscapes. Steps are taken to conserve buildings of high architectural merit or historical importance. But the new approach must be broader than this. It can be realised within the present plan-making procedures. It should take account of the growth of public opinion in favour of conserving the familiar and cherished local scene. It should also have care for the conservation of existing communities and the social fabric, wherever public opinion points clearly towards it. Conservation of the character of cities ought more strongly to influence planners at all stages of their work; conservation of the character of cities should be the starting point for thought about the extent of redevelopment needs; and conservation of the character of cities should be the framework for planning both the scale and the pace of urban change.

Thus 'conservation' had now billowed out to embrace the conservation of 'existing communities' and 'the social fabric'. This particular

The Shambles, Manchester: preserved in the midst of redevelopment.

form of rhetoric has not reappeared in later government pronouncements on the subject. Lord Sandford's statement marked a temporary effusion of conservation sentiment, reflecting the reaction against the passion for 'comprehensive redevelopment' of the 1960s.

Circular 46/73 was prompted by the work being done by local planning authorities in preparing the new type of development plans introduced in the Town and Country Planning Act 1968 and the consolidation Act of 1971.[4] It advised that 'the new structure plans must include adequate policies for conservation in this wider sense as well as in respect of Conservation Areas designated under section 277/1971.' This new advice was also needed because the structure of local government had been comprehensively reorganized in 1973 and the new local authorities were getting ready for their first full year of operation in 1974. The new concept of conservation as described by Lord Sandford, therefore, while unprecedented in scope, was to be brought firmly within the new development plan system created by the 1968 Act, although this was not something that the Planning Advisory Group itself had contemplated.

In February 1974 the Prime Minister, Edward Heath, called a general election which led to the return of a Labour government under Harold Wilson. In view of its very slender majority, Wilson called another general election in September which resulted in the Labour government that lasted until May 1979. Conservation policy, however, survived these political upheavals.

In July 1974, the DOE produced a further circular on *Historic Buildings and Conservation* (102/74). It reminded the new local authorities of their responsibilities for listed buildings, and the 'growing public concern that the national architectural heritage was being rapidly eroded.' It invited local authorities to bring to notice any buildings not yet listed but which 'appear to them to be of list quality', and urged them to make more frequent use of their power to make building preservation orders on buildings not listed but which they considered to be of special importance. This was all the more necessary because the Department recognized that some lists had not been fully revised since the 1950s, and that the current re-survey programme would take another fifteen years.

Circular 102/74, however, also introduced a cautionary note that had not been expressed in quite these terms in previous policy guidance: 'It is perhaps necessary to emphasise again that listing does not itself imply that the building will necessarily be preserved, but merely requires that the building's case for preservation is examined under the statutory procedures for listed building consent applications.'

The most interesting aspect of circular 102/74 was the section which for the first time addressed the 'overlap' between the listing of historic buildings and the scheduling of Ancient Monuments. The circular explained that 'This duplication may appear confusing' (as it had to the Treasury committee back in 1948 – see chapter 9) 'but there is no legal conflict'. Section 56(1) of the Town and Country Planning Act 1971 provided that the controls over works to listed buildings should not apply to 'those which are scheduled and otherwise protected under the ancient monuments code.' Scheduling meant that the monuments became the responsibility of the 'Secretary of State, whereas 'the protection of listed buildings as an extension of planning control is primarily the responsibility of local authorities.' Nevertheless 'There is, however, an area of overlap in respect of, for example, barns, bridges and guildhalls and more recently, industrial structures; and lines of demarcation cannot readily be determined because of the differences in the respective criteria and the methods of selection.'

Evidently this apparent duplication had been a source of concern to Ministers, but they had 'concluded that all structures of a standard eligible for listing under the Town and Country Planning legislation should continue to be listed.' Some listed buildings not yet scheduled might become so at a later date but this had to be tolerated: 'This procedure will involve some duplication, but it will ensure interim protection for monuments that might later be accepted for scheduling; and it will ensure the maximum protection and should remove the danger of structures falling between the two methods of protection.' Thus the apparent rivalry between the two systems that had existed since 1947 was resolved, with the planning process providing a safety net for ancient monuments.

The circular concluded by quoting excerpts from circulars 53/67, 57/68 and 61/68, thus reflecting the continuity of policy that prevailed despite the change of governments. Finally, the appendix to the circular set out the latest criteria for listing, which had not previously been published. These now included 'a very few selected buildings of 1914 to 1919', and also certain categories of buildings which had previously been included in the Supplementary (non-statutory) lists as Grade III – a category which Ministers had decided to abolish in 1970. It is interesting to compare this appendix with the original *Instructions to Investigators*, extracts from which are reproduced in Appendix B. The 1974 appendix is therefore reproduced in Appendix C. The comparison serves to demonstrate the gradual change in the scope of the lists over the past thirty years or so.

Later in 1974 a further circular (147/74) was published which explained the provisions of the Town and Country Amenities Act 1974.[5] This Act extended demolition control to all unlisted buildings in conservation areas (whereas previously this applied only in areas designated under section 8/1972). The circular also gave further advice on the designation of conservation areas. Evidently it was felt that progress in designation had been too slow, and it was now suggested that it was not necessary to precede designation with 'time consuming surveys and the preparation of a conservation policy', which could be done after designation. This somewhat cavalier attitude might be thought to devalue the conservation area concept and no doubt partly accounted for the rapid increase in the number of conservation areas.

Circular 147/74, however, concluded on a far firmer note: 'This circular suggests that local authorities should adopt marginally different expenditure priorities in the field of conservation from those that are now applying. It should not be read as suggesting in any way that there should be an increase in expenditure on planning services without a proportionate saving elsewhere.' One does not have to be clairvoyant to detect the severe voice of the Treasury in this stark warning. It put a damper on further public expenditure over the next few years but it did not dampen popular enthusiasm for conservation. In fact 1975 had already been designated as European Architectural Heritage Year, and an account of that is given in the next chapter.

Covent Garden

The reaction against 'comprehensive redevelopment' reached its epitome in the case of Covent Garden. This was a place of quite exceptional architectural and historic interest, and it has reflected the changing attitudes towards conservation over a period of some three hundred years – and indeed is still the scene of controversy. The area was a produce garden of Westminster Abbey in the Middle Ages. In 1630 the then owner, the Duke of Bedford, decided to develop it as the first of London's great residential squares, somewhat on the lines of the Place des Vosges in Paris, although it was called 'the Piazza' and was obviously influenced by Italian examples. Inigo Jones was the architect. On two sides were splendid groups of houses with tall arcades and giant pilasters. The south side faced the gardens of Bedford House. On the west side was Inigo Jones's church of St Paul, which was burnt down in 1795 and rebuilt by Thomas Hardwick. Pevsner notes that the rebuilding was done 'very correctly, with all the respect which the Burlingtonians had taught architects to show towards the English Palladio.'[6] Bedford House was demolished in 1703 and all the Inigo Jones houses were demolished and replaced in 1769, 1880 and 1890. Early in the eighteenth century the Duke of Bedford decided to allow the square to be used as an open market and in the nineteenth century large and handsome market buildings were erected which filled most of the square.

The fruit and vegetable market continued until it was relocated to Vauxhall in 1974. The Greater London Council then proposed a massive commercial redevelopment of the whole area (except St Paul's Church). These proposals aroused furious opposition from local residents and conservationists. The government eventually responded to this furore when the Secretary of State for the Environment, Geoffrey Rippon, ordered a revision of the statutory list for the area, which resulted in a large number of previously unlisted buildings being added to the list – in effect a comprehensive application of 'spot listing'. As a result the redevelopment scheme was abandoned and the empty market buildings were refurbished and became a hugely popular tourist attraction. The story is not yet complete, however, since in the 1980s the Board of the nearby Covent Garden Opera House produced proposals for enlarging their building which would have required the demolition of some of the remaining eighteenth-century houses: that dispute is not yet fully resolved. Thus Covent Garden mirrors the course of conservation and redevelopment in London over the past three centuries.

An indication of changing attitudes towards conservation and the increasing efficiency of the planning system's protective power can be found in Hermione Hobhouse's book *Lost London*, first published in 1971.[7] After illustrating some of the most grievous losses to London's architectural heritage that had occurred during the twentieth century, and particularly in the post-war years, she ended by identifying five buildings that she feared were in imminent danger. These were Rules restaurant in Covent Garden, Coutts Bank in the Strand, Highgate Cemetery, New Scotland Yard and the Albert Bridge. In fact, more than twenty years later, none of these buildings has been lost. Rules restaurant is still flourishing and has benefited from the conservation of the Covent Garden area. Coutts Bank has been rebuilt but to a fine design and as part of the restoration of the terrace of which it forms the centre-piece. Highgate Cemetery has been rediscovered and is being restored by volunteers. The police have left New Scotland Yard but it has been converted for use as offices for Members of Parliament. Albert Bridge has been repaired and restored. There have been notable losses elsewhere, often the late Victorian and Edwardian buildings that had not yet been listed. But it would be difficult to point to any listed buildings of major importance that have been lost since 1971.

NOTES

1. Cmnd. 4506.
2. Draper, P. (1976) Creating the DOE. *Civil Service Studies*, Civil Service College. See also Radcliffe, J. (1991) *The Reorganisation of British Central Government*. Aldershot: Dartmouth Publishing.
3. 2 Eliz c.52.
4. 2 Eliz c.72 and 2 Eliz c.78.
5. 2 Eliz c.32.
6. Pevsner, N. (1957) *The Buildings of England; London, the Cities of London and Westminster*. London: Penguin, pp. 301–303.
7. Hobhouse, H (1971) *Lost London*. London: (revised edition 1976).

15

European Architectural Heritage Year

For many of those involved in the work, 1975 marked a high point in the post-war history of conservation. European Architectural Heritage Year (which soon acquired the infelicitous acronym EAHY) was intended by its originators to focus European attention on conservation and to raise its profile.[1]

Subsequently some came to feel that it raised the profile so high that thereafter it had nowhere to go but down. Writing in 1978 Alan Dobby refers to 1975 as 'the euphoric time before conservation in Britain was jeopardised by greater than usual economic problems.' He reckoned that it marked the high spot of British conservation in the 1970s but expressed some anxiety that it might be followed by 'a counter-reaction with the reassertion of change in the environment.'[2] In the event, while there may have been a certain sense of anti-climax after the heady days of 1975, conservation continued on its way and, if anything, gathered momentum over the next two decades.

On closer inspection of the records, one is bound to feel that EAHY, at least as it was conducted in Britain, had something of the character of a village fete. The Great and the Good, as patrons of the event, were there in numbers for the opening and closing ceremonies. The organization was elaborate, the amount of money raised was modest, and a large number of small activities were generated. The government, after making a nominal monetary contribution, stood well back and most of the detailed work was done by local enthusiasts.

In the aftermath one is left to wonder what all the fuss was about. But it was generally felt to have been worthwhile, and it stimulated a longer-lasting interest in conservation at the grass roots level. It was perhaps the beginning of the populist concern for conservation which increasingly supplanted the elitist tradition of conservation in Britain.

The Origins of EAHY

EAHY did not emerge suddenly in 1975 like the special issue of postage stamps that accompanied it in Britain. It was the fruit of assiduous cultivation over a long period and three years careful preparation. This was necessary if it was to be a truly European event backed by national governments.

UNESCO started the ball rolling in 1964 when it founded the International Council of Monuments and Sites (ICOMOS) to promote the international study of conservation and to encourage its spread to less developed countries that had a rich archaeological heritage. This generated a series of conferences, at one of which, in 1969 at Brussels, Lord Duncan Sandys first canvassed the idea that became EAHY. The conference set up another body, the Intergovernmental Committee of Monuments and Sites (not the same as ICOMOS; this was a time when international bodies proliferated). This committee called for a 'charter' on conservation, which in turn led to the initiation of EAHY, starting at Zurich in 1971 and culminating in 1975. Lord Duncan Sandys was chairman of the International Organising Committee, which included representatives of twenty-four participating countries, the Assembly of the Council of Europe, the Commission of European Communities, UNESCO, ICOMOS, the European Travel Commission and Europa Nostra (the last of these was founded in 1963 by the Consultative Assembly of the Council of Europe, with Lord Duncan Sandys again as chairman).

How this unwieldy body ever succeeded in launching EAHY is a mystery, but the UK responded by setting up an elaborate structure of committees and working parties to oversee the national programme.

Objectives

In May 1972 the Council of Europe's Committee of Ministers formally endorsed the EAHY concept and designated 1975 for that purpose. The objectives were defined as follows:

To awaken the interest of the European peoples in their common architectural heritage;

to protect and enhance buildings and areas of architectural or historic interest;

to conserve the character of old towns and villages;

to assure for ancient buildings a living role in contemporary society.

The DOE issued a circular in August 1972 (86/72) urging local authorities to prepare schemes that might merit an EAHY award. It was suggested that these might include 'the preparation of townscape and conservation studies, measures to reduce traffic and parking in architecturally sensitive areas, the creation of pedestrian precincts, careful selection and siting of street furniture and lighting, and the rehabilitation of ground surfaces. The restoration of old buildings, and the removal of unsightly advertisements may all contribute to enhancing the national architectural heritage'. Thus 'conservation' became extended, in terms of government policy, to a wide range of environmental improvements of the kind which the Civic Trust had been assiduously promoting for the past fifteen years.

Organization

On 1 August 1972 the government announced that HRH the Duke of Edinburgh had consented to become President of the UK Council for EAHY and that the Civic Trust had been appointed as the UK Secretariat to organise the British programme. The UK Council included four Secretaries of State as Vice-Presidents, with an Executive Committee chaired by the Countess of Dartford and fourteen members including Lord Duncan Sandys, the Duke of Grafton and Viscount Muirshiel. There were about 114 other Council Members, starting with the Archbishop of Canterbury, the Archbishop of Westminster and the Chief Rabbi, plus Sir John Betjeman, half a dozen other luminaries and representatives of over ninety national organizations and institutions. It was the QUANGO to end all QUANGOs. Whatever its purpose was, it met only three times. There were separate committees for Scotland, Wales and Northern Ireland; a Business and Industry Panel, a Craft Skills Working Group, an Education Panel and yet more Panels for Film and Television, Museums, Grants, Tours and Events, and for Youth.

Grants

Over the period 1972–76 the government made an initial grant of £50,000, later increased to £177,750, to the Civic Trust to help meet its costs in administering the campaign. In terms of the sheer number of activities and events it seems to have got value for money.

In addition to these running-cost grants, the government allocated £150,000 for Heritage Year Grants under Section 10 of the Town and Country Planning (Amendment) Act 1972. This type of grant was intended for conservation schemes in conservation areas designated as 'outstanding' by the Historic Buildings Council. In 1972, there were 347 such areas out of a total of some 4,000 Conservation Areas. The Heritage Year grants were awarded by a panel set up by the Executive Council and chaired by Mr. A.A. Wood (at the time Chief Planning Officer of the West Midlands Metropolitan County Council).

The panel distributed 25,000 copies of a leaflet inviting grant applications. Some 1895 applications were received and 129 separate projects were offered grants. A full list of the projects that went ahead was published in the final report of the UK Council. They were mainly modest in size: in the hundreds rather than thousands of pounds. They ranged from £14 for the restoration of an 'old milestone' in the village of Settle, North Yorkshire, to £100 for removing stumps of diseased elms from the village green at Chenies, Buckinghamshire; and topped by £10,000 to the City of Bolton for environmental improvements at Firwood Field. Among grants made for the conservation of individual buildings were £1,250 for the 'restoration of chapel for use by brass band' at West Wycombe, Berks, and £125 for the restoration of a pump house at Great Budworth in Cheshire. The vast majority of grants were not for building conservation, however, but for minor works of environmental improvement – relaying cobbles, undergrounding electricity cables, tree surgery, landscaping, pedestrianization, riverside walks and so forth.

The significance of this grants programme lay not in the individual schemes but in the fact that government money was being made available not simply for buildings of acknowledged importance but for a vast range of minor environmental improvements. Henceforth 'conservation' acquired a much broader meaning and extended to almost anything that could be said to 'preserve or enhance' a conservation area. Thus EAHY was used to advance the cause of conservation in this wide sense and helped to stimulate popular support and ever-increasing demands for conservation works designed to protect and improve neighbourhood amenity.

Other Activities

In keeping with this broad-based popular approach, the UK Councils promoted educational publications, wall charts, exhibitions, competitions, conferences and other activities. The national village fete was in full swing.

At the international level, the Council of Europe identified 51 pilot projects intended to demonstrate various aspects of conservation. The UK attracted four of these grants. One enabled Chester to carry forward the conservation work proposed in one of the 1964 Four Towns Studies (see chapter 12). Two grants went to the historic centre of Poole and to Edinburgh New Town, where in both cases long-term conservation schemes were already in progress. The fourth grant was for the 'Little Houses Improvement Scheme' sponsored by the National Trust for Scotland, focused on workmen's and fishermen's cottages and the houses of small merchants. The last of these schemes again extended the scope of conservation work beyond the grander type of properties that attracted grants from the Historic Buildings Council, and showed that much humbler vernacular buildings could also warrant care and protection.

Support from the private sector was prolific but without the egregious commercialism and self-promotion that such an event would require in the 1990s. Coca Cola sponsored an award scheme for young people's conservation projects. Penguin Books marked the completion of Pevsner's *Buildings of England* with an exhibition. Rank-Xerox sponsored five issues of a special European Heritage journal in five languages. There was a spate of 'Heritage' books, notably from the Architectural Press whose new titles included a special issue of the *Architectural Review* on *Architectural Conservation in Europe* edited by Sherban Cantacuzino, who also produced a report on *New Uses for Old Buildings*. Thames & Hudson published *The Destruction of the Country House 1875–1975* by Roy Strong and others to coincide with an exhibition at the Victorian & Albert museum. The London Region of the RIBA, with support from the Blue Circle Group, published *A Guide to Selected Buildings in London of the 1914–1939 Period*, which demonstrated that conservation was already catching up with the inter-war years. The BBC published *Spirit of the Age* to accompany its TV series. Many local authorities published new county and town guides, together with marking out numerous 'town trails' in places from Fife to Kensington.

Souvenirs flooded the market, including the UK secretariat's own line of posters, lapel badges and envelope stickers. Harrods introduced a line of men's ties with architectural motifs. Staffordshire County Council produced heritage tea towels, jotter pads, windscreen stickers and plastic carrier bags – all to be replicated everywhere in future years. The Department of the Environment struck a set of five bronze medallions showing buildings in the Department's care, only to be upstaged by the Civic Trust's issue of twenty five silver medals featuring some of the nation's principal architectural treasures. Part of the proceeds from the latter were to go to the new Architectural Heritage Fund (see below). One can sense that it was all great fun and a good time was had by all.

Architectural Heritage Fund

One of the principal aims of the UK campaign was to establish a national Architectural Heritage Fund to provide loan capital to local preservation trusts to conserve historic buildings. The idea was that these trusts would acquire and repair decaying buildings, find new uses for them and sell them, recycling receipts into the Fund. The National Trust for Scotland's 'Little Houses Scheme' operated on this basis. Thirty to forty such trusts were set up, some initiated by county councils with starting capital. Overall about £480,000 was raised for this purpose during the year, and it was hoped to bring this to £1 million. The Secretary of State for the Environment promised matching funds, and the government were to nominate half the members of the management committee. It was intended that the fund should start operations at the beginning of the financial year 1976–77.

What is Our Heritage?

The Department of the Environment marked the end of EAHY by publishing a book compiled by Lady Dartmouth entitled *What is our Heritage?*[3] She responded to the question posed by the title in her introductory paragraph 'Is it the mills of the North, the half-timbered houses of the Midlands, or the thatched cottages of the South? Is it the palace, the church, the railway station, the water tower, the manor or the little village green? It is all of them.'

The book then illustrated a large selection of conservation projects completed during Heritage Year, focusing especially on the conversion of old buildings to new uses. These included Samlesbury Hall, a fifteenth century black and white house in Lancashire, converted to a theatre and craft workshops; Telford's Ivory Warehouse in St. Katherine's Dock, London (1820) converted to flats; the latrines at New College Oxford (1386) converted to a new lecture hall; the Regency Terrace at Buxton, Derby, converted to a public library; a Georgian prison and police station in Bath, converted to twelve flats. There were new pedestrianization schemes at Old Harlow, Essex; York Minster; Commercial Street, Leeds; Thames Street, Dorchester; and examples of landscaping, cleaning and floodlighting projects. Indicative of the growing interest in the industrial heritage were the Perth waterworks converted to a tourist centre; Castle Mill at Dorking, Surrey, converted to a private house; and the basement cheese store in a Southwark warehouse converted to a staff canteen.

It is a remarkable anthology of creative conservation work, and justified the Countess of Dartford in expressing her hope that 'we Europeans can preserve and improve our legacy from the past and merge it successfully with our plans and dreams for the future.' The Duke of Edinburgh provided a somewhat sombre introduction to the book in which he said 'Judging from the indiscriminate destruction of old buildings and their replacement by socially and aesthetically disastrous structures in recent years, it suggests a quite remarkable degree of arrogance for the lessons painfully learnt in previous generations. The great excitement of EAHY has been to draw attention to the shortcomings of our generation as curators of the European architectural collection.' One senses the characteristic tones of HRH the Prince of Wales in this message from his father.

EAHY Concludes

The biggest event of the year was the Congress of Amsterdam which attracted over 1,000 delegates, including Heads of State, and ran from 21–23 October 1975. The UK sent the largest delegation of over 100 participants. The most noteworthy feature for our purposes was that a series of working sessions dealt with 'Conservation in the context of urban and regional planning'. At this stage, planning and conservation were seen as inter-related – a message which, as we have seen in chapter 11, the Ministry of Housing and Local Government had been promulgating since the 1960s.

Despite the seemingly overweight superstructure of national committees and nominal patronage, and despite the small scale of most of the projects generated, and even smaller scale of government financial support, EAHY must be accounted a success. It was long remembered by those who took part in it. It demonstrated the wide scope of conservation and confirmed its wide public support and hence its political significance.

SAVE

One other notable event in the history of conservation occurred in 1975, though not directly related to EAHY. This was the founding of SAVE Britain's Architectural Heritage. It was set up by a small group of architectural writers and conservation activists with no resources other than their own knowledge and enthusiasm: 'We each had to put in a tenner', as Marcus Binney recorded later.[4] Binney, now President of SAVE, was at that time architectural writer for *Country Life* and was evidently the most active member of the group. Its concern was with those buildings that had fallen into decay, or were rapidly reaching that stage, and which had failed to attract the interest of the authorities or any of the available sources of public funds for conservation. SAVE set out to show that almost any historic building could be rescued, given sufficient tough determination and imagination, and that alternative uses could be found for them. As Marcus Binney wrote in 1975 'Practical, and sympathetic, solutions can be found for the great majority of historic buildings if only those who decide their fate can be persuaded to give them a chance.'[5]

Ten years after SAVE's foundation, Marcus Binney published an account of some of their heroic achievements (and a few of their failures) in *Our Vanishing Heritage*.[6] Ten years later another set of battle trophies could be displayed. It is an astonishing story and those who are tempted to consign a neglected, derelict historic building to demolition, or to give up hope of restoring it and finding a new use for it, will have SAVE ready and waiting to pounce. They do not always win but they have shown what can be done – and they are still going strong, twenty years after EAHY.

NOTES

1. Civic Trust (1975) *Final Report of the UK Council*. London: Civic Trust.
2. Dobby, A. (1978) *Conservation and Planning*. London: Hutchinson.
3. Department of the Environment (1976) *What us Our Heritage?* London: HMSO.
4. Binney, M. (1995) Interviewed by Charles Darwent. *Perspectives*, May.
5. Binney, M. (1985) *Our Vanishing Heritage*. London: Arlington Books.
6. *Ibid*. p. 9.

PART 3
CHURCHES

16

Churches

At this point it is convenient to break the chronological sequence of events that has been followed so far, in order to deal with the subject of churches in a single chapter.[1]

This chapter deals primarily with churches owned by the Church of England, which warrant separate treatment for two reasons. Firstly, the Church of England has a vast heritage of historic buildings. In evidence to the Environment Committee in 1986 (see chapter 20), the Church's representations recorded that the Church of England owned 16,700 churches, of which 8,500 were pre-reformation and 12,500 were statutorily listed, including 2,675 Grade I. Secondly, the Church has always had its own elaborate legal system and procedures for dealing with the building, alteration and demolition of churches.

Over the past hundred years or so the Church has moved from a position where it insisted on remaining exempt from the evolving statutory system of preservation that applied to other historic buildings, to one where it gradually came to accept an increasing measure of State control in exchange for financial assistance, while still maintaining a degree of independence. The result has been a long sequence of legislation, both ecclesiastical and Parliamentary, which interacts with and is supplemented by various non-statutory agreements or understandings. The legal history is complex and this chapter deals with it only in outline.

Ecclesiastical Exemption

From the time of the first Ancient Monuments legislation in 1882, the church authorities resisted any attempt by the State to include churches in the system of protection. The Ancient Monuments Consolidation and Amendment Act 1913 formalized this 'ecclesiastical exemption' by excluding 'ecclesiastical buildings in use'. Those that had become disused could be scheduled but could not be taken into guardianship or attract State aid.

The Town and Country Planning Act 1932, which introduced the power to make building preservation orders, also excluded ecclesiastical buildings in use. The Town and Country Planning Acts of 1944 and 1947, however, did not exclude such buildings from the statutory listing process, and they became subject to normal planning control. But when listed building control was introduced in 1968, ecclesiastical buildings in use were excluded. Thus alterations and extensions affecting the external appearance of a church required planning permission but not listed building consent. As a result of subsequent changes in planning law, demolition of a church no longer in use requires consent if it is listed or is not listed but is in a conservation area.

So far as Church of England procedures are concerned, alterations and repairs to churches in use are now governed by the Inspection of Churches Measures approved by Parliament, and by the Faculty Jurisdiction exercised by Chancellors and Archdeacons with the advice of the Diocesan Advisory Committees. The process is overseen by the Central Council for the Care of Churches. Redundant churches are subject to other procedures as explained later.

The precise extent of ecclesiastical exemption, as regards the denominations concerned, the types of ecclesiastical buildings and the types of work involved, was for many years indistinct. Thanks to a settlement reached in 1994, these questions have been largely resolved. But it took many years to achieve this and it is necessary to trace that history in some detail.

Church and State

Two problems gradually increased and forced the Church of England to come to terms with the State concerning its architectural heritage: first the huge expense of maintenance and repair, and secondly the related problem of redundancy. There have been two attitudes in the Church towards these combined problems. The first gives priority to pastoral needs and is strictly utilitarian. It regards church buildings as necessary equipment or 'plant' which is required only to meet the current needs of the Church and which, when rendered redundant by changing needs or falling congregations, can be dispensed with. Allied to this approach is the view that the Church should not spend its limited financial resources on costly maintenance when they are required to improve the stipends of the clergy and to build new churches where needed. Against this severely practical view is that which regards almost any church as an object of reverence, a consecrated place and a silent witness to the faith, which is in a deep sense never redundant and should never be neglected or demolished. Those who value church buildings for their architectural and historic importance see no reason why they should be protected any less carefully than other historic buildings. One of the Church's problems has been to strike a balance between these conflicting views and to reach a decision in each particular case. For the same reasons, the State has needed to be circumspect in intervening, but could not be wholly indifferent to this large part of the national heritage.

Alterations

We have seen in chapter 2 how in Victorian times different problems arose from the drastic activities of the church restorers who, while preserving many mediaeval churches from ruin, also ruined many by their radical rebuilding and doctrinaire views on ecclesiastical architecture. This was not a new problem for the Church. As long ago as 1237 a Constitution of Otho, the Legate in England of Pope Gregory IX, made in a national synod, strictly forbade 'Rectors of churches to pull down ancient consecrated churches without the consent and licence of the bishop of the diocese under the pretence of raising a more ample and fair fabric.'

This measure of 1237 is said to be the origin of the system of jurisdiction whereby the Church's Consistory Courts regulated alterations to an existing church or the demolition of one no longer needed. The Church itself was satisfied with this antique system and strongly defended its own jurisdiction when that came to be questioned by those who were concerned that it did not afford adequate protection to church architecture. When the issue was raised during the passage of the Ancient Monuments Consolidation and Amendment Act 1913, Archbishop Davidson defended the Consistory Courts and assured the House of Lords that the Archbishop of York and he believed 'speaking largely, that the authority which at present controls these matters is the authority which can best control them in the years to come.'

Archbishop Davidson did, however, undertake to commission an enquiry to see whether the existing system afforded sufficient protection or should be improved. This led to an inquiry conducted by Sir Lewis Dibdin, Dean of the Arches, whose report in 1914 resulted in the setting up of Diocesan Advisory Committees in each diocese to provide expert advice to the Consistory Courts on repairs, redecoration and other alterations. In 1918 the Central Council for the Care of Churches was established as a source of research and advice.

Redundancy

In the years prior to World War II attention was focused on the need to control ill-advised restoration and alteration. But a greater problem was the growing number of churches that had become disused. An elaborate procedure had evolved since the time of Henry VIII for dealing with this problem, but as this often involved the unification of two or more parishes, the result was usually that at least one church became redundant. Schemes of reorganization could provide for such churches to be demolished or put to other use. Under legislation of 1923, proposals to demolish a church had to be laid before both Houses of Parliament for two months, and the advice of the Royal Fine Art Commission had to be sought on measures to protect churches of 'archaeological, historic or artistic interest'. None of this ritual, however, could avert the problems of redundancy or secure funds for preservation.

In 1948 the Church Assembly appointed a committee chaired by the Bishop of Norwich 'to investigate the problem of disused and unwanted churches'. This committee found that there were about 400 churches which were seldom or never used, of which 300 were thought to be of architectural or historic interest. Their report recommended that churches of sufficient architectural or historic importance should be offered to the Ministry of Works for taking into guardianship under the Ancient Monuments Acts. Discussions were held with the Ministry and a list of 111 potential candidates was drawn up. But nothing came of this proposal.

Archbishops' Commission on Redundant Churches

From 1948 to 1958 some 230 churches were demolished, many as a result of war damage or comprehensive redevelopment. The problems of redundancy were becoming even more apparent, and in 1958 the Archbishops of Canterbury and of York set up a new commission to advise on redundant churches. This was chaired by Lord Bridges, recently retired as Permanent Secretary of the Treasury. There were seven other members, appointed in a personal not representative capacity, and these included the Permanent Secretaries of both MHLG and MOW, and the renowned archaeologist Sir Mortimer Wheeler. It was a

Westminster Abbey before alterations to the West front: print by Hollar (1652).

committee of laymen, apart from the Bishop of Grimsby. (I was one of the two joint secretaries for the first eight months, my colleague being from the Church Commissioners.) The commission took evidence from a wide range of interests, including the Friends of Friendless Churches, which was always prepared to defend the friendless church even when all other measures had been exhausted (it still exists).

The Archbishops' Commission's report was published in 1960 and provides a very clear and succinct account of the problem and the labyrinthine procedures for dealing with it that existed at that time. The report began with the observation that 'All the evidence which we have received shows that the review which we have been asked to undertake is a matter of great and urgent concern, both within the Church and for the nation at large.'

Statements were sought from the diocesan bishops which showed that some 370 churches were at present redundant and that another 410 were expected to become redundant within the next twenty years. According to MHLG's lists, 440 of the total 780 churches were of special architectural or historic interest, and a further 86 were of some interest. The geographical distribution of redundancy was uneven, being worst in declining city centres and in sparsely populated rural areas. Some dioceses had very few redundant churches, but five each had fifty or more. An appendix to the report showed that there were 14,222 parishes, of which 3,431 were rural, 3,641 urban and 7,150 'mixed'. If anything, the report probably substantially underestimated the extent of redundancy existing at that time and likely to occur in the future.

The report set out in considerable detail the

Westminster Abbey with Hawksmoor's west towers (1735): print from Maitland's *History of London* (1739).

existing procedures for dealing with redundant churches and for authorizing their demolition or measures for keeping them in repair. The commission concluded:

It is clear to us that a new and improved procedure is called for. Little damage may have been done so far; but the position is very dangerous and we are convinced that irretrievable damage will be done during the next few years unless, first, a procedure can be established under which the claims of churches to preservation are taken into account early in discussions of reorganisation, and, secondly, funds are assured for preserving those which have to be declared redundant but which nevertheless ought to be preserved.

The Commission's recommendations fell under two main heads. First, a new simplified procedure was proposed for declaring redundancy, including reference to a new Advisory Board to assess the case for preservation. The board was to be appointed by the two Archbishops after consultation with the Prime Minister, which seems an odd burden to add to his other duties but which reflected the fact that the problem of redundant churches was seen to affect both Church and State. Secondly, there should be a Redundant Churches Fund, established by statute, in which redundant churches would be vested and to which the

Church, State and public at large would contribute.

It seemed likely that the proposed fund would need to take responsibility for between 300 and 400 churches over the next fifteen years. The capital cost of putting those into reasonable repair was put at not less than £1½ million to £2 million spread over twenty years, with an eventual total annual maintenance cost of between £60,000 and £70,000 at 1960 prices (the 1995 equivalent would be at least twenty times as much and would still look to be a gross underestimate).

Over the next few years, the government and the Church took steps that largely implemented the Commission's recommendations. The Church reformed its procedures for declaring a church redundant and set up a Central Advisory Board for Redundant Churches to which proposals for demolition had to be referred for advice on architectural value. In 1968 the Church established the Redundant Churches Fund, whose object was 'the preservation, in the interests of the nation and of the Church of England, of churches and parts of churches of historic and archaeological interest or architectural quality', being those vested in the Fund. The Fund told the Environment Committee in 1986 that it 'understands "preservation" in broad terms to mean the minimum care necessary to keep a vested church and its contents in their state as they existed at the time of the vesting, but in good repair.' The fund was financed 60 per cent by the Department of the Environment and 40 per cent by the Church on a quinquennial basis: the funding for 1984–89 was £6m (with provision for inflation). At that time the fund held 202 churches and expected to receive twelve more each year and to repair twenty; there were seventy repair schemes in progress. The fund submits an annual report to the Advisory Board for Redundant Churches and to the Church Commissioners who transmit it to Parliament.

Faculty Jurisdiction Commission

Thus arrangements for redundant churches became well established but negotiations between Church and State on what to do about churches still in use took longer. The Church wanted to retain ecclesiastical exemption while at the same time needing financial assistance with repair. The government was reluctant to agree to both. Negotiations began in 1971 and in 1975 the government announced that it accepted in principle the case for some contribution from the State. This led to the setting up in 1977 of the Scheme of State Aid, which was originally intended to run for a trial period of five years. It was a condition of this agreement that the Church would not move to demolish a listed church (or a non-listed one in a Conservation Area) without first notifying the Secretary of State for the Environment and allowing time, if required, for a non-statutory public inquiry. The purpose of this interim agreement was intended to allow time for the Secretary of State to review the whole question of ecclesiastical exemption.

The Church likewise set in train a review of its own procedures and in 1980 appointed the Faculty Jurisdiction Commission, whose terms of reference were to review both the law of the Church of England and the secular planning legislation as far as it affected the Church. The Commission's report published in 1984 made 230 recommendations, mostly of a detailed procedural and technical character, including improvements in consultation and accountability, and a revised constitution for the Cathedrals Advisory Commission.[2] There was nothing very radical in these proposals: they were aimed at simplifying and clarifying the existing procedures, and at establishing a new system for cathedrals.

The Commission addressed the question of ecclesiastical exemption in some detail, noting that 'there have been indications of a demand for reconsideration of the case for the exemption and the question of its continuance has become an issue of public importance.' After marshalling various practical arguments in favour of retention, however, the Commission concluded that:

> The heart of the problem lies much more in the distinction which we recognise between a purely preservationist form of control imposed from the outside and a means of guiding architectural or decorative work in the service of the Church so that the expression of its own identity through its buildings comes naturally from within. The former method can only be concerned with the protection of what exists whereas the latter, which is available through the faculty jurisdiction, can unite care of that kind with a creative architectural approach which we see as 'celebratory'.

Finally, the Commission concluded that 'the exemption of churches in use from listed building control is sound in principle; and that, subject to certain reforms in the faculty jurisdiction, its continuance would be beneficial both to the Church and to the wider community.'

One member of the Commission, however, disagreed fundamentally with his colleagues. This was Marcus Binney, who was appointed to the Commission when it was half way through its work. He was architectural editor of *Country Life* and chairman of Save Britain's Heritage (see chapter 15). His minority report was printed with the Commission's report. He began by declaring candidly that 'My view at the outset was that the so-called ecclesiastical exemption should be ended, and that the Church of England should be subject to listed building control. Slightly to my own surprise, virtually everything I have heard and read as a member of the Commission has strengthened this view.' Evidently he failed to persuade his colleagues to his point of view.

Surprisingly, the Commission's terms of reference did not extend to redundant churches but they addressed the subject briefly in their report and concluded that

> It does appear to us that there could be advantage to the Church as a whole if at the final stage (of the existing redundancy procedures) after all the possibilities under the Pastoral Measure had been exhausted, and when the Commissioners had decided that there was no alternative but to demolish, the exemption were lifted and the building in question brought under the normal listed building control.

The advantages to the Church were seen as relieving it of 'some of the odium which often accompanies demolition and of the expense of maintaining unwanted buildings.' In actual fact, listing would not necessarily confer either of these advantages; rather the opposite. But there was the possibility that if listed building consent to demolition were refused, the Church might be able to serve a purchase notice on the local authority requiring them to acquire the building.

ENVIRONMENT COMMITTEE 1986

Despite the Commission's willingness to allow redundant churches to come under listed building control (after the Church's own procedures had been exhausted), the established system remained in place pending the Secretary of State's review. No decision had been taken by the time of the House of Commons Environment Committee's enquiry in 1986.

The Environment Committee was told in evidence that in the seventeen years since the Pastoral Measure came into operation, the future of 1,053 redundant churches had been decided. Three-quarters of them were still standing and over half of these (583) had been found new uses. Two hundred had been transferred to the Redundant Churches Fund, and 266 churches that did not warrant transfer to the fund had been demolished, of which 62 were listed buildings; but in none of these cases did the advisory board or the local planning authority object. The Church felt that this record justified both the continuation of ecclesiastical exemption and an increase in financial help from the State.

The Environment Committee did not touch on this subject in its recommendations, except to recommend that grants should be extended to cathedrals, which had until then been excluded both from the normal procedures and from grant-aid. The Church had proposed (as a 'quid pro quo', as it said in its evidence) that in future works affecting cathedrals should be referred to a new Commission of Review, comprising three members, one of whom would be appointed by the Secretary of State. In its response to the Environment Committee the government simply noted that English Heritage already had power to give grants to cathedrals but that this was a matter for them: 'It is question of priorities'.

Lord Skelmersdale's Statement

Negotiations continued between the government and the Church authorities, together with the Churches Main Committee, representing other churches and religious organisations. This led eventually to a statement by Lord Skelmersdale, the Parliamentary Under Secretary of State in the DOE, during the Committee Stage of the Housing and Planning Bill in 1986.[3] This also took account of a judicial decision in the House of Lords (Attorney General ex-relator Bedfordshire CC v. the Trustees of the Howard United Reform Church, Bedford), which held that the existing form of ecclesiastical exemption did not apply to the total demolition of a church. The main points of Lord Skelmersdale's statement were that listed building consent or conservation area consent would in future be required for demolition of a church, except in the case of the Church of England where demolition was consequent on a pastoral or redundancy scheme. For other churches consent would be required where demolition would materially affect the architectural or historic interest 'such as a spire, tower or cupola'. The Church of England would, however, always ask the Secretary of State whether he wished to hold a non-statutory public inquiry in those cases where the Historic Buildings and Monuments Commission, the Advisory Board for Redundant Churches, the local planning authority or a national amenity society gave 'reasoned objections' to the proposed demolition. In considering his recommendation following such an inquiry, the Secretary of State would 'take into account the financial implications of retaining a church building as well as the architectural and historic interest of the church and other planning and social factors' (such provisos have not been applied specifically to any other type of listed building, although the costs of retention may be just as significant for the owner). Since planning control applied to all alterations that materially affected the external appearance of a church, listed building control was unnecessary (again, this did not apply to other types of listed building where both planning and listed building consent were required).

Thus the Church of England retained (for the time being) the ecclesiastical exemption that it had traditionally enjoyed, but that was now made more explicit and subject to 'voluntary arrangements' regarding consultation with the Secretary of State.

The Wilding Report

As part of the Secretary of State's protracted review of ecclesiastical exemption, the Department of the Environment and the Church Commissioners in July 1989 appointed Mr Richard Wilding (formerly Permanent Secretary of the Office of Arts and Libraries) to carry out a review of workings of the Redundant Churches Fund. His report was published in March 1990.[4]

Wilding found that during the first twenty years of the Fund's existence 1,261 redundant churches had been dealt with under the Pastoral Measures, and that of these 696 (55 per cent) had been appropriated to an alternative use, 297 (24 per cent) had been demolished and 168 (21 per cent) had been preserved. At the end of 1987 there were 67 more which had been declared redundant but whose future had not yet been decided. That number of redundancies was far higher than nad been forecast by the Bridges Commission in 1960, when it was estimated that 370 churches were then redundant plus another 410 likely to become so within the next twenty years. Of the 696 churches for which alternative use had been found, the largest group (235) were adapted for 'civic, cultural or community use'. 146 were residential and the remainder in a variety of commercial and other uses, including 88 used for 'worship by other Christian bodies' (including the Roman Catholic church).

Since 1977 the DOE and English Heritage had been making grants for the repair of churches in use, and in the past twelve years grants had been made to 3,500 churches of 'outstanding' architectural and historic importance. In many of those cases the grants may have averted redundancy, and only about 25 had been declared redundant.

From 1969 to 1989 the government had contributed about £12.35m to the Redundant Churches Fund and the Church had found some £12.7m from the sale or lease of redundant churches and sites; for the quinquennium 1989–94, however, the government undertook to contribute 70 per cent of the total budgeted expenditure and the Church 30 per cent.

Wilding concluded that the system was well organized and represented a 'solid record', but that it could 'work still better'. He went on to observe, however, that 'The preservation of redundant churches is not self-evidently a good thing'; he asked 'What purposes does it serve?' After a sensitive assessment of the varied interests involved, he asked 'Would the interests of both Church and State be served by a slow but inexorable accumulation of every redundant church, wherever it may be, for which no suitable alternative use can be found and which has some feature of architectural, archaeological or historical interest?' He concluded that 'These are questions of policy which are plainly for the Government and the Church authorities, and ultimately Parliament and the General Synod to answer.' (One might note that somewhat similar questions may arise in relation to other types of listed building.)

Meanwhile, however, Wilding concluded that the present system should continue, with some improvements to pull together more effectively the various interests involved, and that the Advisory Board and the Redundant Churches Fund should not be merged (as had been suggested by some), but that the RCF should be renamed more accurately as 'The Churches Heritage Trust' (it has since been renamed the Churches Conservation Trust). Finally, he recommended that the DOE, the Church Commissioners and the RCF should be jointly responsible for the 'preparation, agreement, carrying through and monitoring of a financial and operating plan' for dealing with redundant churches on a triennial basis. That recommendation has not yet been acted on.

Wilding also raised the possibility of 'secularizing' the whole process, relieving the

Church of any continuing responsibility for redundant churches and treating the ecclesiastical heritage essentially as one aspect of conservation policy. A new body, 'something rather like an ecclesiastical cousin of English Heritage', would become responsible for building up and maintaining a representative but selective collection of English church architecture. In the longer term the new body might be merged with English Heritage. Wilding concluded, however, that this radical and intriguing 'secularization solution' would be regarded by the great majority of those concerned as a 'sad mistake'. He shared that view himself.

The Wilding Report provides much the most detailed and thoughtful discussion of the problem since that of the Bridges Commission. Although he shied away from the more radical alternatives, his report raised issues which, as he said, were 'not yet pressing but will have to be answered one day.'

Cathedrals

Cathedrals, which had previously been treated as a separate class due to their special form of governance, and which enjoyed greater freedom from control, are now subject to controls under the Care of Cathedrals Measure 1990, the care of Cathedrals Rules 1990 and the care of Cathedrals (Supplementary provisions) Measure 1994. All of which should be sufficient to deter the amateur from interfering.

Until 1990 cathedrals received no government grant, but in that year English Heritage was allocated an extra £11.5m (over three years) to set up the Cathedrals Repairs Grants Scheme and instituted a survey of all 61 Church of England and Roman Catholic cathedrals to assess repair needs over the next decade. By 1994–95 a total of £15 million had been offered to 54 cathedrals.

Future Prospects

Two years after the publication of the Wilding Report the Secretary of State for the Environment, with the Secretary of State for Wales, announced that an Order would be made to provide that ecclesiastical exemption would in future apply only to the Church of England and to other denominations and faiths which set up acceptable internal systems of control embodying the principles contained in the government's code of practice.[5] That code is set out in full in DOE Planning Policy Guidance PPG 15 of September 1994,[6] supplemented by a booklet published by English Heritage.[7] It requires full publicity and consultation with the local congregation as well as with those bodies normally consulted, plus provision for announcing and recording the decision reached, and arrangements to ensure the proper maintenance of historic churches, including thorough inspections on a fixed cycle of not more than five years.

The Order redefines ecclesiastical exemption on a more limited basis, and all places of worship not included in it are fully subject to listed building and conservation area control. Control now also extends to internal fixtures, furnishings and the 'arrangement' of interiors. Those concerned are advised that 'It is particularly important to identify, and where possible retain, the spatial arrangements and fixtures that belong to the principal period of building' (such injunctions are not normally applied to other types of listed building).

Buildings no longer in ecclesiastical use are

not covered by the exemption and are subject to normal planning and listed building control. The courts have decreed that a church building that has been in use but is about to be demolished is not in ecclesiastical use for the purpose of these procedures.

Thus, after more than a hundred years of ecclesiastical exemption, the Church of England (and churches of other denominations whose inclusion had been uncertain) are now in effect subject to at least as restrictive a form of control as other listed buildings (although the procedures differ) and in some respects even tighter control (e.g. as regards interiors) than applies to other types of listed building. This has been the price of accepting financial aid from the State, and in recognition of public concern.

The policies now set out in PPG 15 seem to have been generally welcomed, but they will not in themselves solve the problems of redundancy or those of proper care and maintenance of churches still in use. Firstly, the huge financial crisis that confronts the Church of England, following its period of disastrous property investment, means that the Church can no longer afford to pay its share of the cost. The Church Commissioners have proposed that no more than three churches a year should be transferred to the Redundant Churches Fund, though they have also decided to set up a 'temporary maintenance account' to help meet the costs of maintaining churches whose future is still uncertain. Secondly, experience has shown (as has been apparent for many years) that the Church's own system for controlling interior alterations to church buildings can all too easily be avoided or ignored, and there are no effective sanctions to prevent this, apart perhaps from the Archdeacon's displeasure. The present system is still so protracted and uncertain that delay and decay can render long-term preservation impossible. The ecclesiastical exemption leaves wide scope for misconceived internal alterations that are not caught by planning control and may escape the surveillance of the Church's Advisory Committee. For these and other reasons many of those informed and concerned about the heritage of church architecture have little confidence in the present system. It may well be that within the next decade a quite different system will be needed to ensure that the government's policy on this aspect of the heritage is effectively implemented and that public concern is allayed.

NOTES

1. The principal sources for this chapter are: *Report of the Archbishops' Commission on Redundant Churches 1958–1960*. London: SPCK, 1960. Environment Committee: First Report Session 1986–97, 146–1. London: HMSO, 1987. (See memoranda nos. 18, 19, 22 and Appendices 13 and 43–50). See also chapter 20.
2. *The Continuing Care of Churches and Cathedrals* (Report of a Faculty Jurisdiction Commission). London: C10 Publishing, 1984.
3. Lord Skelmersdale's statement was explained in detail in *DOE Circular 8/87*, paras 203–5. London: HMSO, March, 1987.
4. *The Care of Redundant Churches* (The Wilding Report). London: HMSO, 1990.
5. *The Ecclesiastical Exemption (Listed Buildings and Conservation Areas) Order, 1994* made by the Secretary of State for the National Heritage.
6. Department of the Environment (1994) *Planning Policy Guidance: Planning and the Historic Environment, PPG15*. London: HMSO.
7. English Heritage and Cdw (Welsh Historic Monuments) (1994) *The Ecclesiastical Exemption: What it is and How it Works*. London: English Heritage.

PART 4
1976–1995

17

PRELUDE

European Architectural Heritage Year in 1975 (see chapter 15) stimulated a great deal of interest and activity at the local level, and did much to strengthen the growing public concern for conservation. But there was still a good deal of ambivalence among the planning profession in attitudes towards conservation, and a similar uncertainty about the government's attitude, despite the excellent work done by MHLG during the 1960s.

At the start of EAHY the Royal Town Planning Institute's journal *The Planner* ran a special issue on planning and conservation.[1] Timothy Cantrel, who was to head the UK secretariat for EAHY, commented that, despite progress in policy and legislation, 'it can still be said that planning undervalues conservation.' But he predicted that 'conservation will graduate from being a small and specialised sector of planning to being a new approach to planning itself.' In the same issue Gordon Cherry, the planning historian, remarked that the exponents of conservation were 'a minority group who drag a by-and-large uninterested community in their wake.'[2]

Over a year later, the *Architects' Journal* ran an editorial 'Looking at the Listing Process' in which it expressed concern at the Department of the Environment's recent proposal to delegate to local authorities all applications to alter Grade II buildings, except where they had received government grant for conservation (the Historic Buildings Council had expressed similar concern).[3] The *Architects' Journal* noted that in the years 1969–75 there had been over 11,000 applications to alter Grade II buildings, 'So there are grounds for considerable foreboding . . . at what appears to be the thin end of the wedge as far as DOE's shedding of responsibility for listed buildings is concerned.' Their concern was aggravated by a speech that Lady Birk, a junior Minister in the DOE, had made the previous week in which, the *AJ* claimed, she had said that 'too many buildings have been listed' and that in future 'we shall not be giving so many buildings the benefit of the doubt.'

This report of Lady Birk's alleged comments produced a reply from her two weeks later in which she claimed that what she had said was that 'I think in the past we have been listing too many *marginal* buildings, particularly those of the later nineteenth century' – a response that was hardly likely to reassure the *AJ*, and was singularly ill-judged in view of the growing interest in buildings of that period and the numbers lost during the town centre redevelopment boom of the 1960s.[4] She also disclosed that she had discussed with Sir John Summerson, Chairman of the Advisory Committee on Listing 'whether we could find a way of being more specific (as regards listing nineteenth century buildings) without being too exclusive.' But evidently Sir John had declined to assist in this way and Lady Birk had to be satisfied with an assurance that Investigators would be asked 'to give close attention to these marginal cases (which she estimated at

200–300 a year), by looking at them more critically' and to get a second opinion from another Investigator and from the Historic Buildings Council. No other Minister had ventured to question the listing process or to cast doubt on the government's commitment to conservation: rather the opposite. So Lady Birk deserves a small place in our history of the subject.

The DOE failed to follow MHLG's practice of publishing an annual report, which was a loss to historians, and to the Department itself since it meant that there was no ready source of reference to aid the fallible Departmental memory. But the course of conservation over the next few years can be traced through the HBC's Annual Reports.[5] Although those reports concentrate on HBC's activities, they include brief comments which provide some indication of the tide of opinion.

This period began with the death of Lord Holford in 1976. He had been one of the most prominent figures in the 'listing' business for thirty years, although his own professional career was chiefly concerned with architecture and planning rather than conservation. He was not a scholar, unlike his colleagues John Summerson, Nikolaus Pevsner and others on the Advisory Committee.

In its report for 1975–76 the HBC included a separate section on Conservation and Planning. It noted the 'ever-hardening public reaction against wholesale redevelopment', and that the Secretary of State (Mr Peter Walker) had declared that 'the bulldozer is being pensioned off.' The HBC warned that 'Historic towns and villages, city centres and suburbs, will survive only if conservation is accepted as a major planning objective.' It claimed, which was not entirely true, that 'In this country conservation has always been part of the planning process, both in towns and in the countryside;' and added that 'Even in those towns not generally regarded as "historic", sensible planning can do much to preserve and enhance their individual character.' Venturing somewhat beyond its normal field (and impinging perhaps on the Royal Fine Art Commission's patch) they observed that 'The matter of design is more delicate . . . There is a feeling that there are opportunities for matching but not copying the vernacular of the past which would avoid the visual disasters to be seen in many historic towns.'

In its report for 1976–77 the Council took up these themes again: 'Conservation of the historic environment cannot be effective in isolation. It must be treated as part of the planning process for the environment as a whole, and other policies must be coherently guided to this end. This has been the view of the Council in recent years and 1976–7 saw a welcome, though as yet not complete, acceptance of its implications.'

By 1977 the worsening economic climate began to affect the HBC's work. Its funds, which it considered to be very inadequate, were increased only by the amount needed to offset inflation in building costs, and the staff engaged on listing was cut. Nevertheless, there were now 246,000 listed buildings in England and 16,423 had been added during the year, including seventeen Grade I. No Grade I buildings had been demolished in 1976 and only 104 Grade II, compared to 351 in 1975. Ninety Town Schemes were in operation, although London had only one such scheme – for the Rugby estate in Camden.

In concluding its report for 1976–77 the Council for the first time paid tribute to the work of local authorities: 'In the past, it was the Government which had to take the lead, while local government and local opinion was hesitant or opposed. Now that the tide has turned, perhaps due in no small measure to Government efforts, we must not be found lacking in the effectiveness of our response.'

In its report for 1977–78 the Council said 'We believe that the listing procedures which have been evolved over a number of years are an essential part of the planning process.' It was odd that the Council felt it necessary to say so but it shows that the relationship between planning and conservation was still not taken for granted.

1978–79 marked the twenty-fifth anniversary of the HBC. The Council was pleased to note the fact that its scope had been steadily extended but was concerned that its budget had not been increased proportionately. It must have wondered what the future held in store for it when the General Election of October 1979 resulted in the return of a Conservative government. It was to find out two years later.

NOTES

1. Cantrel, T. (1975) *The Planner*, **61**(1).
2. Cherry, G.E. (1975) *The Planner*, **61**(1).
3. *Architects' Journal*, 13 October 1976.
4. Birk, A. (1976) *Architects' Journal*, 27 October.
5. Historic Buildings Council *Annual Reports*. London: HMSO.

18
English Heritage is Born

A Consultation Paper

In November 1981 the Conservative government published a consultation paper *Organisation of Ancient Monuments and Historic Buildings in England.*[1] It was an early example of what would come to be recognized as the classic Thatcherite style. It deserves to be remembered on that account as much as for the proposals themselves.

It began by noting that the government currently spent about £36m a year on ancient monuments and historic buildings in England, and that there were now some 275,000 listed buildings and 12,500 scheduled monuments. In 1980 two million people visited the 400 monuments in the care of the DOE but income from entrance charges, sale of souvenirs etc., totalled only £7.5m.

Here, clearly, was an under-utilized resource that cried out for a more business-like approach. The consultation paper struck the new chord in its fourth paragraph: 'The presentation of monuments to the public' was to be seen as 'a significant commercial operation.' Change was needed 'to bring more professional expertise to the promotional and commercial side of the ancient monuments operation.' Moreover, 'the Government considers that a more imaginative approach to promoting monuments could lead to much more income being generated.' This was a language unfamiliar to conservationists. It could hardly have been less calculated to win their support.

The consultation paper noted that the DOE itself employed over 1,000 people on this work, and that the Royal Commission on Historical Monuments (England) had a further 100 staff. The Ancient Monuments Board for England had eighteen Board members but no staff of its own, and the Historic Buildings Council for England likewise had twenty members but no staff; both these bodies were serviced by the DOE. The government's view was that this was a self-contained block of work which 'does not in itself require to be carried out by Civil Servants.' It therefore proposed to transfer the work to a new government agency, together with the existing staff who would thereby magically cease to be civil servants and no longer be included in the total civil service manpower count, thus advancing the government's commitment to make major cuts in the number of civil servants.

The government considered that this 'more imaginative approach' needed 'commercial and entrepreneurial flair.' Anticipating the likely reaction of the conservationist lobby to these proposals, the consultation paper affirmed that 'The Government does not see the aim of improving the commercial performance of ancient monuments as conflicting with the basic priority to preserve and protect the monuments for future generations.' So, in the Brave New World, ancient monuments were expected to improve their performance.

While most of these Departmental functions were to be transferred to a new agency, the Secretary of State would retain various statutory functions, including decisions on listing and scheduling, on taking monuments into care, standards of maintenance, and policy on the opening of monuments to the public. In addition 'There is no intention to transfer any of the Secretary of State's planning functions' to a new agency. 'The broadest aspects of policy in this field would of course remain the responsibility of the Secretary of State', while the agency 'would be charged with the implementation of those policies.'

It was also proposed to amalgamate the Ancient Monuments Board and the Historic Buildings Council as part of the new agency. Thus, as the consultation paper pointed out, 'The net result of the changes would be to reduce the number of non-Departmental public bodies by one.' This was in the days when the government had determined radically to reduce the number of Quangos (quasi-autonomous non-governmental organizations) and before they decided greatly to increase them to take over the running of hospitals, schools, etc. The government had not yet decided whether the Royal Commission on Historical Monuments should be amalgamated with the new body, and views were sought on that.

Subject to the outcome of this consultation, the government proposed to introduce the necessary legislation in the 1982–83 session, so that the new agency should start work in April 1983. The government expressed the hope that the new agency would 'command greater respect in the heritage field' than the DOE (a somewhat gratuitous insult to those concerned in the Department). Finally, not only could the visitor 'expect better value for money' but, rather surprisingly, 'the monuments would gain increased prestige and importance.'

The reaction to these proposals, especially from the various long-established conservationist bodies, was one of incredulity, contempt and anxiety.

The Way Forward

Having digested the response to the consultation paper, the government produced its final proposals the following year in a booklet *The Way Forward.*

Ministers had clearly been somewhat chastened by the ferocity of the response. This was hardly surprising in view of the naive, almost crude, way in which its proposals had been presented. More than three hundred organizations and individuals replied to the consultation paper, including all the leading lights of the conservation movement and many other bodies with wider interests. The government evidently felt obliged to reply to these comments in a more conciliatory manner. Although it had no intention of withdrawing from its main proposal, it sought to deal in more detail with 'some of the uncertainties and misunderstandings' that the consultation paper had evoked.

In a section headed 'Major Worries', the government declared that it did 'care intensely about the heritage' and that it had no wish to 'distance' itself from heritage work. There was no mention now of 'the commercial performance of ancient monuments'. Instead the government 'believed that the potential of our built heritage for the education and enjoyment of young people has received far too little attention up till now': indeed, this aspect had not been mentioned in the original consultation paper.

The Department apparently felt somewhat bruised by the fierce reaction provoked by the consultation paper and replied plaintively that 'people were expecting more from the paper, and reading more into its omissions, than was ever intended.' After reviewing the general objections raised, however, 'The Government's conclusion is that there is a clear balance of advantage in England in the creation of a new public sector agency, on the broad lines described in the consultation document', and that this 'is in the best interests of the heritage.'

The booklet then went on to itemize the specific points and matters of detail that had been raised by those knowledgeable about the subject, but in nearly every case these were said to be matters that would require further consideration, including the precise division of functions between the DOE and the new agency, the composition of the Board, the allocation of staff, the advisory and executive functions, research, pay and conditions, and so forth.

Turning to more weighty matters: 'One particularly important decision remains to be taken, namely on the title of the new agency.' This was clearly just the kind of thing that requires the mature judgement of Ministers and their Permanent Secretaries. While not yet ready to take a final decision on this issue, 'In the Government's view the words "national" and "heritage" should be avoided.' The reasons for this strange policy decision were that the word 'national' would 'generate confusion', and the word 'heritage' had 'wider connotations than the functions which would be assigned to the new body.' The name proposed was the 'Commission for Ancient Monuments and Historic Buildings (England)', but the government, perhaps forgetting its earlier reservations, also decided that it should be known for short as 'English Heritage'. Ten years later it went further and established the Department of National Heritage, which became responsible for English Heritage as well as for sport and the National Lottery, together with other miscellaneous functions. Thus the word 'heritage' did indeed prove to have wider connotations.

National Heritage Act 1983

The Bill that established English Heritage was introduced in the House of Lords on 25 November 1982. Opening the debate the Earl of Avon, as parliamentary Secretary at the Department of the Environment, referred at the outset to the uproar that the proposals had initially provoked:

> Some people have talked about this proposal in terms of selling the heritage for a profit. Nothing could be further from the government's mind. This Government have, I believe, an impeccable record in terms of protection of the heritage.

In the event, none of the eighteen peers who spoke in the debate opposed the main proposal to set up a Commission for Historic Buildings and Ancient Monuments (HBMC) and to transfer to it most of the relevant functions carried out by the DOE. As Baroness Birk, leading for the Opposition, said 'I do not believe that this Bill will divide the House on party lines.' She paid tribute to the Secretary of State, Mr Heseltine, 'who has shown tremendous concern with the heritage and also a very productive interest in these matters.' She thought that the new arrangements would 'give both freedom and inspiration to the safeguarding of our buildings.' She ended by saying that, provided there was adequate funding, 'it could result in a new and healthier prospect for our ancient monuments and historic buildings.'

Evidently a great deal of effort had been expended by Ministers to disarm the conservationist bodies and other critics who had originally bitterly opposed what they thought to be the government's intentions. That opposition was no doubt stimulated mainly by the sheer vulgarity of Thatcherite language in which the proposals had been expressed in the original consultation document. But in the protracted consultations that followed, those chiefly concerned had been reconciled to the idea of a new independent agency with specific responsibility for conservation. It was clearly felt that such a body would be able to represent conservationist interests more effectively to government and also, perhaps, be more open to lobbying, than when that function was merely one among many Departmental responsibilities.

The Duke of Grafton, who had been a member of the Historic Buildings Council since its inception thirty years ago, and also chairman of the SPAB for a similar period, remarked on 'the complete change in public opinion' over those thirty years. 'I think that people have simply become not prepared to lose their heritage of well-loved buildings, which before 1953 were disappearing at a most alarming rate.'

The greater part of the Bill dealt with changes in the administration of various national museums, the Armouries and the Royal Botanical Gardens at Kew. As Lady Birk said, it was really two Bills in one and much of the debate focused on the provisions relating to museums, in which several of their Lordships had a personal interest as Trustees. The HBMC did not make its appearance until Clause 30, where its functions were set out in somewhat bald terms:

(1) The Commission shall have the functions conferred on them by any enactment, and shall exercise them for the purpose of securing, so far as practicable, the conservation in England of Ancient Monuments, archaeological areas, and buildings and areas of special architectural or historic interest.

(2) So far as practicable, the Commission shall provide in England education and instruction to the public in relation to Ancient Monuments and buildings and areas of special architectural or historic interest.

By the time the Bill had completed its passage through both Houses, the Commission's functions had been revised in Section 32 to read as follows (while still subject to the otiose caveat 'so far as practicable'):

(*a*) to secure the preservation of ancient monuments and historic buildings situated in England;

(*b*) to promote the preservation and enhancement of conservation areas situated in England; and

(*c*) to promote the public's enjoyment of, and advance their knowledge of, Ancient Monuments and historic buildings in England, and their preservation.

In addition, the Commission were to exercise certain Ministerial functions concerning ancient monuments and historic buildings. Those functions occupied twelve pages in Schedule 4 and included the care and 'presentation' of some 400 properties previously managed by the Secretary of State and the making of grants for conservation. The Commission also became responsible for advising the Secretary of State on scheduling and listing, and on applications for listed building consent. The Secretary of State, however, remained responsible for general policy matters and for all decisions affecting private property (scheduling, listing, preservation orders). He also retained all his functions in relation to normal planning control where those impinged on ancient monuments and listed buildings.

The Bill then moved on to Committee Stage and their lordships did their usual conscientious job in examining the Bill line by line. In particular they worried about the proposed name of the new Commission, whose full acronym would be CFAMAHBFE. Lady Birk suggested 'Heritage Commission for England', which was not far short of 'English Heritage', but this important matter was not resolved in the Lords. One other point of interest was the amendment initiated by Lady Birk, which introduced a reference to 'Town and country planning' into the list of attributes to which the Secretary of State should have regard in appointing members of the Commission, so that this finally read 'having knowledge or experience of one or more of the following: namely history, archaeology, architecture, the preservation or conservation of monuments or buildings, town and country planning, tourism, commerce and finance.' The fact that knowledge or experience of planning was not originally thought to be useful in the work of the Commission says something about the gap that still existed between conservationists and planners.

The government introduced the amendment spelling out the Commission's functions in more detail, and this was welcomed. Lady Birk tried to get 'and conservation' added to 'preservation', which led to an inconclusive discussion about the distinction between the two terms. Lady Birk sensibly said that, as the two were virtually, but not quite, synonymous, it would be better to include both – but this was not accepted at this stage. However, when the Bill reached its Third Reading, Lady Birk again put down an amendment to add 'conservation' to 'preservation', which gave the House yet another opportunity to chew over this puzzling point. But discussion was curtailed by Lord Avon's announcement that the government now accepted that the two words had different connotations and that the Bill should include both – though without distinguishing between them.

When the Bill, as amended in the Lords, moved on to the House of Commons, Mr Heseltine was no longer responsible for his baby as he was now Secretary of State for Defence. The job fell to Mr Paul Channon as Minister for the Arts. He opened the Second Reading debate on 24 February 1983 by observing that 'Debates in another place (i.e. the House of Lords) and elsewhere have shown a wide measure of agreement for the view that the new body should produce a more positive and creative approach arising from the fusion of existing arrangements. I hope that it will exert a powerful influence on conservation in its widest sense.'

For the Opposition, Mr David Clark (MP for South Shields) said 'We do not wholly oppose the Bill. We think that it has considerable potential. But it has many shortcomings', though he did not say what these were. Mr Toby Jessel (MP for Twickenham) pointedly asked why the new body had to be 'dynamic', when it was responsible for preservation? To this Mr Clark replied 'because in the past we have not shared our culture and our heritage as we should . . . I believe that dynamism is essential.' As an example of what he meant by dynamism he spent much of his speech advocating the 'reconstruction' of ancient monuments, which must have come as a surprise to most of the Bill's supporters.

On the perennial question of what the new body should be called, Mr Christopher Murphy (MP for Welwyn and Hatfield) proposed 'English Heritage Commission'; but for the government Mr Giles Shaw, Under Secretary of State at the DOE, replied that 'A title that includes the confusing word "heritage" would cause confusion.'

At Committee stage (22 March 1983) Mr Cormack for the Opposition proposed endowing the new body with £10 million as a start-up fund. This was opposed by the government but was agreed to by the Committee by nine votes to eight (presumably some government supporters were AWOL). This generous proposal was, however, defeated at Report stage. The Bill finally received Royal Assent on 13 May 1983, with no further substantial amendments, and the Appointed Day for its coming into operation was left to be determined by Order. The new body retained its statutory name as the Historic Buildings and Monuments Commission for England but soon adopted the shorter name that the government had earlier suggested – English Heritage.

NOTES

1. Department of the Environment (1981) *Organisation of Ancient Monuments & Historic Buildings in England: A Consultation Paper.* London: HMSO.

2. Department of the Environment (1982) *Organisation of Ancient Monuments and Historic Buildings in England: The Way Forward.* London: HMSO.

19

English Heritage at Work

First Annual Report

English Heritage (referred to here as 'the Commission') came into being on 1 October 1983 but did not assume its full responsibilities until 1 April 1984. Its parent Act required it to produce an annual report and accounts, and its first report covered the eighteen months of its existence. It was a glossy full-colour production, very different from the traditional HBC format. Similar annual reports have been published each year since then. They give a very full and confident account of the Commission's work and it is not necessary to duplicate that here. A brief summary of its main activities can be given but it is more relevant to our purpose to see whether the reports reflect any changing policies or attitudes towards the heritage.

In his introduction to the first annual report, the Commission's Chairman, Lord Montagu of Beaulieu, gave expression to its new-found sense of independence and purpose. He spoke of the Commission's 'wider objectives' which included 'the assertion of the distinct identity of the Commission' and its 'duty to speak out clearly for the preservation of our heritage' – as it had done 'in numerous public debates, including the implications for historic buildings of the Value Added Tax changes proposed in the 1984 budget, the abolition of the Greater London Council and the fate of the notable buildings in its care and controversial urban developments in the City of London and elsewhere.' This was a new and powerful voice for conservation.

In its first full year the Commission received £52 million in government funding and raised a further £7.4 million in receipts from admissions and sales. There were ambitious plans for expansion but the Secretary of State 'thought that the pace at which we wanted to move was too fast' because of the demand on resources.

English Heritage inherited about 400 historic buildings and archaeological sites from the DOE, together with responsibility for their care and maintenance. These accounted for £18.2 million expenditure in 1985–86 – about 32 per cent of the total budget. These properties now attracted some four million visitors a year and English Heritage were quick to recognise both their 'marketing' potential and the need for better presentation. A membership scheme was introduced, similar to that of the National Trust, providing free entry to sites, a newsletter and events for members. By 1994 total membership had grown to 310,000 (still far below the National Trust, which had over two million members). Dover Castle was singled out as a major attraction; consultants proposed a ten year plan designed to increase visitors to nearly 400,000 a year and to increase annual income from admission charges and sales to over £1 million.

As regards casework, the Commission advised the Secretary of State on 667 applications for works to scheduled monuments but expressed the view that this 'double handling' of cases was inefficient and they urged that responsibility for decisions on such cases should be transferred to the Commission. On historic buildings, local planning authorities were required to refer to the Secretary of State proposals, which they were minded to approve, involving the demolition of any listed building or alterations to Grade I or II* buildings. The Secretary of State was required to refer such cases to the Commission for advice. In 1894–85 3,200 cases were referred and the Commission's advice was accepted in all but two cases. But the Commission did not at this stage propose that the decision on such cases should be transferred to them.

During 1984–85 the Commission offered 169 conservation grants totalling £7.2 million for outstanding secular buildings, 486 grants totalling £3.9 million for outstanding churches, and 800 grants totalling £4 million for building and environmental schemes in conservation areas. The largest single grant offered was £500,000 towards the repair and preservation of Clevedon Pier (whose whereabouts was perhaps not universally known).

On the 'listing' front, the Commission was less closely involved since the 're-survey' launched by Mr Heseltine in October 1982 was being undertaken by County Councils and consultants rather than by the DOE Investigators as was the case with the original lists. This delegation of the work had resulted in nearly 33,000 more buildings being added to the lists in 1984–85 and a further 1,000 as a result of spot-listing (i.e. outside the normal listing programme and in response to threats of imminent demolition or alteration). The Commission noted that 'The re-survey has unearthed a wealth of architectural riches, Devon, for instance, turned out to have significantly higher numbers of buildings of special architectural or historic importance than had been expected. But other areas, notably Humberside, Lincolnshire and much of North Yorkshire, produced fewer.' Interestingly, it had been found that 'some building types have turned out to be less rare than had previously been thought.' But it was not clear whether this had resulted in some buildings being de-listed or simply in fewer buildings of those types being added to the lists, or perhaps merely a modified view of their scarcity value.

A notable feature of this first report by English Heritage was that it paid far more attention to ancient monuments than to historic buildings and gardens: four chapters on the former and only one on the latter. The reason for this was probably that initially the Commission was more concerned about properties in its care (mainly ancient monuments and relatively few historic buildings), and saw its role chiefly in preserving and 'presenting' those properties rather than in dealing with the far greater number of listed buildings for which it had only an advisory role (aside from its grant-making function). In this sense, at least at this early stage of its operation, the Commission appeared to be exercising a less broad responsibility for conservation than had been hoped or as was envisaged in its general publicity. But this balance was progressively redressed in later years as English Heritage exerted its influence more widely in the field of conservation.

Increasing Workload

In its next annual report, for 1985–86, the Commission's work on historic buildings and conservation areas assumed a much more prominent place. The relative priorities were reversed: the report began with historic buildings and areas, running to ten pages, followed by ancient monuments with six. The Commission had perhaps discovered that its functions relating to historic buildings generated far more work than ancient monuments, and that the former offered it far wider scope than the latter. In the first two years of its existence over 61,500 buildings had been added to the statutory lists, bringing the total to 373,000, and the Commission now expected the total to reach nearly 500,000 by the time the re-survey was complete (it had been said that Mr Heseltine set the DOE the 'objective' of listing 500,000 buildings, which seems so fatuous that it may not be true). In 1985–86 the Commission had commented 'positively' (*sic*) on 375 of the 1,074 applications referred to them by local authorities for consent to demolish or alter Grade I or II* buildings. It seems most probable that all such applications related to alterations or partial demolition (e.g. of later additions) rather than to complete demolition. In addition, the Commission advised the Secretary of State to call in 59 of the 3,294 applications for demolition of Grade II buildings referred to him by local authorities who were minded to give consent. But by far the largest quantity of casework arose from the requirement that local authorities should notify the Commission of all planning applications affecting the setting of Grade I or II* buildings and those affecting the most important conservation areas. Since 1 October 1984 this had resulted in more than 27,000 planning applications being referred to the Commission. It seems astonishing that so many proposals should have been made affecting this relatively small proportion of the total number of listed buildings. On the other hand, comparatively few of these cases prompted the Commission to comment on them: 961 in 1985–86.

In April 1986 the load on the Commission was further increased by the transfer to it of the responsibilities for deciding listed building consent applications that were dealt with by the Greater London Council before its abolition. This meant that in London the Commission dealt with all the listed building work that elsewhere in the country was dealt with by the local planning authorities. Not surprisingly, within a few years the Commission was relieved to see this work transferred to the London Boroughs, although that move was opposed by those who did not trust the Boroughs to do the job properly.

An Advocacy Role

Aside from the casework chronicle that occupied much of the annual report for 1985–86, some incidental comments in the report showed that the Commission was developing a new role as part of the conservationist lobby. It had participated in the Joint Committee of Amenity Societies (a voluntary grouping), and had increased its membership scheme to nearly 45,000 and aimed to have at least 80,000 members by the end of the next year. A potential ambiguity was developing between the Commission's role as an agent of government policy and its tendency to ally itself with the conservationist pressure groups. Thus a somewhat anomalous situation arose where the largest and best resourced conservation lobbyist was financed almost entirely by the government itself.

As a result of its explicit concentration on conservation, the Commission also posed a danger that 'conservation' would again be seen as in conflict with 'planning', since the planning function inevitably would often involve the resolution of conflicts between conservation and other interests. When both those functions were dealt with within the DOE, such conflicts had to be resolved within the Department, whereas English Heritage was by definition *parti pris*. Formally, the Secretary of State, not the Commission, was responsible for deciding appeals against refusal of planning permission and listed buildings consent, but he was statutorily required to seek the Commission's advice and the Commission sometimes appeared at planning inquiries to press the conservationist case. There were the seeds here of possible difficulties in the years ahead, as a result of the Commission being endowed with both advisory and advocacy functions relating to the planning process.

The annual report for 1986–87 noted the setting up of a planning branch 'which enhanced our ability to take a more active role in planning issues affecting conservation areas'. The Chairman commented that 'We have been outspoken on conservation issues when necessary but have also argued by word and deed that conservation is not synonymous with fossilisation and must accommodate changing economic and social considerations.' What this might mean in practice remained to be seen.

Before the report for 1986–87 was published, the Commission was called on to give evidence to the House of Commons Environment Committee's enquiry on historic buildings and ancient monuments. This was to be the first time that the heritage had been the subject of a full Parliamentary enquiry. An account of it is given in the next chapter.

English Heritage is by no means the only player on the conservation stage, but its scope and resources obviously make it a major influence. Increasingly it has come to see itself as a leading protagonist and this sometimes brings it into direct confrontation with the government, notably the DOE and the Department of Transport. The Commission takes the view that in this it is acting in pursuit of its statutory duty. The Commissionm has also adopted a similar role in opposing local authority sponsored projects and in opposing private developers at public inquiries on planning appeals. In other cases it stakes out its interest at an earlier stage of the development process. For example, it gave evidence to the Private Bills Committee opposing the plans for development at King's Cross and St Pancras in connection with the Channel Tunnel Rail Link. It intervened in proposals for the redevelopment of the Royal Opera House in Covent Garden, and of Spitalfields Market and of Paternoster Square by St Paul's Cathedral. It entered into negotiations with developers whose development scheme in Southwark uncovered the remains of the Elizabethan Rose Theatre, which led to its preservation beneath the new building. It also intervened in proposals to convert County Hall (vacated on the abolition of the GLC) to hotel use, and in a major development scheme, London Bridge City, opposite the Tower of London. In most of these cases the projects have subsequently stalled or been abandoned, though this may have had more to do with the recession in the property market than with the intervention of English Heritage. Nevertheless, by being prepared to commit itself in these high profile cases, the Commission enhanced its reputation as a guardian of the heritage and added substantial weight to representations made by other conservation interests.

The Commission also plays an important role in commenting on the statutory development plans of local planning authorities and on strategic planning and conservation issues at national level. The 1989–90 annual report observed that 'Good planning policies form the basis of effective conservation', and that therefore the Commission 'tries to influence the planning system in the interests of the historic environment.'

Public Image

The annual reports for 1987–88 and 1988–89 reflected some unease in the Commission about the public perception of its role, and some anxiety and ambivalence in its own mind about the diverse functions it was expected to fulfil. At the conclusion of its 1987–88 report the Commission commented 'We are aware that our undoubted growing success in the leisure market may, in the public mind, have overshadowed our very considerable achievements in the field of conservation.' They returned to this theme in the following year's report, observing that

> It is not always easy to find the right balance between being on the one hand expert practitioners of conservation and on the other participants in the business of marketing the national heritage. Both roles are important and we try to do both well. Yet to the public we may still appear more as part of the leisure market than as the authoritative body in conservation of the historic environment at large.

The next two annual reports did not revert to this theme directly, but they emphasized the more technical, scientific and professional aspects of the Commission's work, its academic and specialist publications, its major role in conservation of its own properties and the growing importance of its educational work. In this its traditional archaeological activities helped to emphasize its seriousness of purpose. On the other hand, the reports also record with evident satisfaction the success of its commercial activities. By 1990 trading income had increased to £9.1m compared to £2.3m in 1984, many new shops had been opened and the range of products sold at them had increased from less than 1,000 to over 9,000. By 1990 the membership scheme had increased to nearly a quarter of a million members, and the Commission's properties were attracting around five million visitors a year. If the Commission felt any unease about its role as part of the heritage and leisure industry, it could comfort itself with the fact that the proceeds from its commercial activities flowed back into its other work for conservation.

Widening Scope

While the Commission's basic work on scheduling ancient monuments and listing historic buildings continued, the scope of that work was continuously extended. These activities included the first comprehensive inventory of archaeological sites (likely to total some 600,000 recorded sites and finds, about 50,000 of which may warrant statutory protection), a national inventory of conservation areas, a register of historic parks and gardens, and a register of battlefields. There was also an extraordinary exercise to list 500 of the K6 or 'Jubilee' type telephone kiosks designed by Giles Gilbert Scott. The Commission having completed this totally otiose task, which made a mockery of listing criteria, the Department of the Environment then announced that the number of kiosks listed would be doubled to 1,000, and the programme was extended to include 216 of type K2 kiosks in London. It must be doubted whether a more absurd undertaking has ever been known in the history of conservation.

Aside from kiosk mania, the Commission increasingly faced demands for the listing of other and more significant building types that had not figured largely in the earlier lists but which came to feature within the ever-increasing scope of 'the heritage'. As the annual report for 1990–91 noted, these included 'types of building which, a few decades ago, nobody would have thought of saving', such as 'mills, warehouses, barns and cinemas' (in fact the earlier lists did include examples of all these

Telephone kiosk, Ladbrook Grove, London, by Giles Gilbert Scott: one of 1200 listed.

but the result was to generate demands for listing many more buildings of these types).

With this burgeoning catalogue of monuments and buildings, the Commission was moved to declare that 'Clearly it is neither possible nor sensible to try to save every old building, structure or monument', and that its task was to identify and recommend 'the most important surviving examples for protection.' In fact, the statutory scope of listing is far wider than this, extending in principle to all buildings of 'special architectural or historic interest', and in practice nearly all surviving buildings prior to 1700 and most eighteenth-century buildings have been listed. The problem of selectivity becomes far more difficult as listing extends into the Victorian period and beyond.

A particular example of this problem has been the question of whether, and if so how, buildings of the 1930s and later should be added to the lists. In 1987 the 'cut-off; date for listing was set at thirty years (i.e. buildings less than thirty years old should not be considered for listing). Following this the Commission recommended 'a number' of twentieth- century buildings for listing, but the Secretary of State for the Environment accepted only eighteen of those recommendations and incorporated these in a selection which the Commission's Advisory Committee considered to be 'unrepresentative of both the building types and the architectural styles of the immediate post-war period.' This decision was probably influenced by the personal views of the then Secretary of State, Nicholas Ridley, who was himself an engineer and the grandson of Lutyens. In 1992, however, the Commission recommended forty seven educational establishments for listing, involving ninety five post-war buildings and all of these were accepted.

This has been a very abbreviated account of the first eight years of English Heritage at work. Throughout that period Lord Montagu of Beaulieu was Chairman. In his last annual report he wrote 'I think I can claim that I hand over to my successor an organisation which is becoming increasingly known and respected by the public for its efforts to ensure that the man-made environment will be conserved and interpreted for future generations to enjoy.' He was surely well justified in that sense of achievement. English Heritage had made a good start, but it was possible that, in asserting such a dominant and extensive role in the field of conservation, it was in danger of assuming responsibilities beyond its resources and perhaps undervaluing the function of local planning authorities.

The arrival of the new Chairman, Mr Jocelyn Stevens, in April 1992 marked a new stage in the story of English Heritage, which is resumed in chapter 21.

NOTE

The references and quotation in this chapter are taken from the *Annual Reports* published by English Heritage.

20

Select Committee

The Committee's Enquiry

In the Session 1986–87 the Environment Committee of the House of Commons undertook an enquiry into historic buildings and ancient monuments. This provided the first opportunity since the Gowers Committee in 1948 (see chapter 9) for a thorough examination of government policy on the heritage. Select Committees of this kind have power 'to send for persons, papers and records', to appoint specialist advisers and to carry out visits of inspection. Their reports are presented to the House and published. The reports include verbatim minutes of oral evidence and copies of the more important written evidence received (all other written evidence is placed in the House of Commons library). Unlike the Gowers Committee, the Select Committee heard evidence in public.

The Committee held hearings on ten days between April and July 1986. Its report was published in January 1987 in three volumes. The report itself (Vol. 1) ran to only fifty pages but the minutes of evidence (Vol. II) occupied 378 pages and the appendices (Vol. III) a further 200 pages. Copious other written evidence was received and placed in the library.

Naturally, the Environment Committee's enquiry was seen by all those bodies involved with the heritage as a unique opportunity to press their interests and display their activities. Their evidence, oral and written, provides a comprehensive account of the state of play, or state of the art, at that time. But the result is somewhat disappointing and certainly did not produce any critical or searching examination of the rationale or philosophy of conservation. The case for conservation was taken for granted by both the Committee and those who appeared before it, and the evidence given concentrated very largely on the 'nuts and bolts' aspects – the effects of VAT and inheritance tax; the detailed procedures for listing; the grants regime and the need for more funds for conservation. In ploughing through the evidence one longs to hear a voice of dissent or unorthodoxy, but the silence is almost deafening. A few episodes, however, provide some interest.

Devil's Advocate

Alone among his colleagues, Mr Sydney Chapman (MP for Chipping Barnet and a professional architect and town planner) attempted to inject a more controversial note into the proceedings and to evoke a more interesting debate. His manner was courteous and somewhat tongue in cheek, but he asked some questions that could have prompted a more stimulating debate. At the Committee's first session of oral evidence he asked the DOE's senior representative (Mr Timothy Hornsby, head of the Directorate of Ancient Monuments and Historic Buildings), 'Do you think there is a danger that in our enthusiasm to list (and I am putting it in a slightly perjorative way) anything we think remotely of historical or architectural interest, we might be defeating the whole object, which should be to conserve our architectural excellence rather than preserve our mediocre buildings?' He pointed out that in 1985, 204 listed buildings were demolished and 2,372 partially demolished. 'To put the question again, do you not think that we ought to have fewer listed buildings and try to save more of them rather than increasing the number, knowing we are going to lose hundreds of them each year?'

Tim Hornsby, however, did not rise to this bait and gave the orthodox response 'The bald answer would be no. Listing *per se* is not an attempt to preserve everything in aspic, it puts up a marker: this building is of architectural and (*sic*) historic interest.' The Committee did not pursue the issue.

Mr Chapman tried again later when the National Trust witnesses were being examined. He asked their chairman, Dame Jennifer Jenkins, 'do I sense that while we would all want to conserve the excellence of our architectural heritage there is a danger that they (i.e. English Heritage) might be trying to preserve too much of the mediocre in our architectural heritage?' Dame Jennifer gave an uncompromising reply: 'I would not agree with that. I think that the criteria for listing are very strict and, of course, there can always be a difference of opinion at the margin. But I think that, although the list will be very large, it includes buildings that certainly deserve preservation.' Again, none of Mr Chapman's colleagues followed up his line of questioning.

Mr Chapman returned to his theme when attention turned to conservation areas. He asked Mr Peter Robshaw of the Civic Trust 'Is there evidence that some conservation areas are far too widely drawn?' Mr Robshaw thought that was possible, and that 'if an area is too big the policies are going to be too generalised to be properly effective.'

Mr Chapman had done his best to be provocative but he admitted that he had been acting as 'devil's advocate' and that at heart he was as much a conservationist as anyone: few, if any, politicians have dared to be otherwise.

The Amenity Bodies

A further opportunity to discuss the scope and purpose of listing and the limits (if any) of preservation, came when the Committee met representatives of the Joint Committee of the National Amenity Societies. Mr Chapman teased them: 'Mrs Sladen, no doubt, representing the Victorian Society will tell me her utopian dream is to see every Victorian building listed and Mr White, representing the Georgian Group, will tell me every building post-1714 should be listed. Do you accept that many Victorian and Georgian buildings must inevitably be demolished?'

In response to this deliberate provocation the witnesses showed admirable restraint. Mr Robshaw for the Civic Trust, said 'I would certainly not assert (that) every listed building has got to be kept.' Mrs Teresa Sladen for the Victorian Society said 'discrimination is extremely important . . . I think it would be quite ridiculous to suggest that every Victorian building or even the majority of Victorian buildings were listed. What is really important is to choose the right ones.' Mr Saunders for the Ancient Monuments Society observed that one in three applications for total demolition succeeded but 'Listing creates the presumption in favour of retention, which can be set aside in the face of clear evidence there is no potential use or there is no public need for the building. It does not canonise a building and say that it must be preserved in perpetuity.'

If the Select Committee's labours achieved nothing else, this brief exchange with the voluntary organizations most concerned with the subject showed that the best informed conservationists were not extremists but held to a sensible and realistic approach. Less well-informed exponents of conservation, however, seldom share those attitudes.

Victorian Buildings

Among the flood of written evidence submitted to the Committee, the most interesting and well considered came from the Victorian Society. It did not share the satisfaction that the other conservation groups expressed with the listing process and its evidence focused sharply on its defects. In its memorandum it noted certain deficiencies. Firstly, the listing criteria appeared to be treated as a constant that should remain unaltered despite changes in historical perspective and architectural scholarship. Although in 1970 the Advisory Committee had identified twenty five Victorian architects whose 'principal works' should be listed, 'Today that list is more interesting for its omissions than for any practical use, and a list of the same kind drawn up today would be similarly flawed as soon as issued.' Secondly, the types of buildings covered by the lists were too limited. Those 'poorly represented' from the 1840–1914 period included 'the suburban villa, the city-centre office buildings, minor public and municipal buildings, most educational buildings (especially the post-1870 Board Schools), and all the more complex social and industrial buildings like hospitals, asylums, work-houses, prisons, factories, baths and orphanages.' Thirdly, and related to the second point, was the fact that it appeared to be harder to get larger buildings listed than smaller ones: examples where listing had been rejected included several Victorian country houses, Longton Town Hall in Stoke-on-Trent, and Essex Hall in Colchester, which was built as a railway hotel and later used as a mental asylum. Only after repeated efforts had it been agreed to list St Bernard's Hospital in Ealing, which had a celebrated place in the history of the treatment of mental illness.

The Victorian Society argued that it appeared that 'practical considerations over re-use, social anxieties and on occasion, political factors' sometimes prevented a building being listed, whereas such considerations should come into play only where the question of demolition or re-use arose and not at the listing stage. Lastly, the Society thought that problems over listing Victorian buildings arose because those responsible for the lists were mostly traditional (i.e. pre-industrial) archaeologists or art historians. The former tended to emphasize the criterion of antiquity, 'while the architects and art-historians have tended to assume that important buildings and objects should be visually attractive.' The Society urged that 'values of technical, social or community interest' should also be included in the assessment for listing purposes. In their view 'the visual criterion, though important, is far from sufficient.'

Much of what the Society had to say represented the leading edge of scholarly conservationist concerns, and raised important questions about a process that had become somewhat complacent. But it evidently passed over the Committee's heads. After hearing its views in oral evidence the Committee Chairman asked 'It is just on listing you are fussed?', and discussion passed on to other matters.

Taken as a whole, the evidence given by the

amenity societies on 7 May provided the most interesting day of the Committee's hearings, but the issues raised did not figure largely in its report.

By far the most voluminous evidence submitted came from the Historic Buildings and Monuments Commission for England (HBMC), whose shorter name of English Heritage was now well established. Its memorandum ran to over fifty pages. While providing a full description of its work and a more detailed account of procedures than the DOE memorandum, it contained little in the way of more general observations on conservation, its scope, purpose and potential conflicts with other objectives. Instead it presented a long list of additional powers and funding which the Commission thought it should have and which it hoped the Select Committee would endorse. This was one way in which it expressed its 'independence' of government and its adopted role as a lobbyist for the conservation interest. Aside from the demand for increased resources (it wanted at least £10m above their present £72m), the Commission proposed that it should take over the Secretary of State's statutory responsibilities for listing and for scheduling ancient monuments. It also asked the Committee to consider the case for the Commission taking over responsibility for the unoccupied Royal Palaces and the work of the Royal Commission on Historical Monuments. It listed nine possible legislative changes, including power for the Commission to issue 'stop notices' to prevent works affecting listed buildings in a conservation area, and power to impose an Article 4 direction (to override the General Development Order) where the local authority had declined to make one. Having recently had transferred to it the responsibilities for listed buildings exercised by the Greater London Council before its abolition, it proposed that it should have the same powers outside London. It also proposed that it should have power to call-in and refuse applications for listed building consent which a local authority was minded to approve, subject to a right of appeal to the Secretary of State. This was a fairly ambitious list of proposals for enhancing its powers from a body that had only been in existence for two years.

The Commission's memorandum ended on a high note:

> We believe that the conservation of the heritage should be regarded as essential in its own right and are committed to that belief. We see it as an essential part of our role to set standards and to act as spokesman for the heritage as a whole. But it would be wrong to see conservation solely as an isolated end in itself. It runs through the well-being of society, it helps to preserve communities; it provides pleasure and education to those who visit our towns and villages, historic buildings and monuments and study our history; it generates employment in the construction industry and in service industries as well; it helps to maintain traditional skills; it earns money through tourism. We believe that the benefits to be derived from support for conservation are immense and that the allocation of resources should recognise those facts.

Other Evidence

Among the seventy memoranda submitted to the Select Committee and printed in its report, those from the Royal Institute of British Architects (RIBA), Royal Town Planning Institute (RTPI), and the Town and Country Planning Association (TCPA) were of significance since those bodies, unlike most of those that submitted evidence, were not part of the conservationist lobby or identified exclusively with conservation interests. The Committee had evidently asked those submitting evidence to address four specific aspects (listing, grants, public access, and effectiveness) and this perhaps limited the scope for any wider observations. In fact these memoranda were confined mainly to detailed aspects of listing, grants and procedures. One has to search for any hint of unorthodoxy or fresh thinking.

Both the RIBA and RTPI urged that the cut-off date for listing should be changed to allow any building over thirty years old to be listed (as was the case in Scotland). The RIBA came up with the suggestion that a 'Grade 4' should be introduced, which could include buildings less than thirty years old but which 'may achieve higher grading in due course'. The RIBA observed that these changes 'would encourage two ideas not always current within the conservation lobby – that some recent buildings may be masterpieces, and that our children's architectural heritage will include buildings designed in our day.' The Institute also commented somewhat ambiguously that 'The purpose of listing is not to prevent all change but to ensure that a building's special character is protected; only the very finest buildings deserve to have a first class ticket to eternity.' But it did not enlarge on this interesting thought or discuss its implications.

The RTPI did not have any notably original thoughts to offer. The TCPA saw scope for improving the listing process and also some scope for simplifying the regime – e.g. by restricting the need for listed building consent to those features (e.g. the facade) that had justified its listing.

The Minister's Evidence

The Committee's final session of oral evidence saw the return of the DOE's representatives together with the Minister of State, Lord Elton. The Minister found it necessary to inject a realistic, somewhat acerbic response to the large claims made by HBMC for increased funds and wider responsibilities. He remarked that, as there were not unlimited funds available, if the Commission wished to spend more on one aspect of its work it would have to do less elsewhere: 'That is a hard choice and I am not at all surprised that they find it difficult and would like to escape it, but it had to be made somewhere.' As to the Commission's bid to take over management of the unoccupied Royal Palaces, 'English Heritage has spent some time explaining to the Committee that they have a great deal to do and insufficient resources to do it. I am not therefore looking in any case for an expansion of their functions.'

The Select Committee's Report

The Select Committee's Report was published in January 1987 as a Parliamentary paper. Its opening sentence declared that 'Historic buildings and ancient monuments are an essential part of the personality of this country. In the course of our inquiry it was repeatedly stressed to us how crucial they are to people's perception of the environment in which they live.' The next sentence, however, reverted to what was to be a principal theme of the report: 'The architectural heritage is also a major tourist attraction.' In a later passage the Committee expressed perhaps its most surprising conclusion: 'We believe that heritage is capable eventually of paying its own way.' In the meantime, however, it recommended that more funds should be made available (unquantified but above the rate of inflation).

So far as English Heritage's role was concerned, the Committee acceded to nearly all of HBMC's proposals and recommended that it should take over from the Secretary of State responsibility for listing and scheduling; that it should have power to serve building

Tudor tavern found inside Docklands warehouse, now relocated and restored as The Dickens Inn (*c*. 1980). (Photo Paul Berkshire)

preservation orders and repair notices; that it should deal with listed building consent appeals and determine applications for works at or near scheduled monuments. In addition there were recommendations for various further grants and tax concessions. Two proposals that had been put to the Committee by various witnesses but which it did not endorse were that the listing and scheduling systems should be combined and that the Royal Commission on Historical Monuments should be transferred to English Heritage.

The Government's Response

It now remained for the government to reply to the Committee's recommendations. Its response was published on 20 January 1988. It began 'we very much welcome the report, not only for its constructive and thought-provoking recommendations . . .; (it provides) a most valuable starting point and a rich quarry of ideas and data for the refinement and development of policy towards the heritage, and is likely to remain valid for this purpose for a considerable time to come.' Cynics might detect a degree of evasion beneath the effusive

Albert Dock, Liverpool: before conversion to Tate Gallery of the North (1984).

tone, and the last sentence indicated that the government hoped that the Committee would not return to the subject for some time to come. But the government claimed that 'well over half' of the recommendations had been accepted in full or subject to certain conditions. One of the Committee's conclusions that could be seized on with enthusiasm was that 'the heritage is of enormous and very often unrealised economic potential.' On the more significant recommendations, the government was not prepared at present to transfer any of the Secretary of State's functions to English Heritage or to promise further grants or tax concessions.

It was a fairly thin haul for the heritage after the 700 pages of evidence, report and response produced by the Select Committee's enquiry.

NOTE

All the references and quotations in this chapter are taken from the published reports of the Environment Committee, Session 1986–87 (London: HMSO, 1987 and 1988) *Historic Buildings and Ancient Monuments*, Volume I, *Report and Proceedings* [HC 146 I] *Historic Buildings and Ancient Monuments*, Volume II, *Minutes of Evidence* [HC 146 II] *Historic Buildings and Ancient Monuments*, Volume III, *Appendices* [HC 146 III] *Historic Buildings and Ancient Monuments, First Report: Observations on the First Report of the Committee in Session 1986–87* [HC 268].

21

Prioritizing the Heritage

The Department of National Heritage

In April 1992 the new Chairman of English Heritage, Mr Jocelyn Stevens, began his term of office. A month later, following the general election, the Department of National Heritage was formed. The new Department covered a strange variety of functions, which seemed to reflect the personal interests of its first Secretary of State rather than any logical grouping. Among its responsibilities for sport, broadcasting and the arts, the new Department took over responsibility for conservation and English Heritage from the Department of the Environment. But it did not take over the DOE's responsibilities for town and country planning, which at once posed problems for demarcation and coordination.

The first circular issued by the Department of National Heritage (DNH) published jointly with the DOE, *Responsibilities for Conservation Policy and Casework* explained that 'policy responsibility for archaeology and conservation of the built heritage' now rested with the Secretary of State for National Heritage, together with 'sponsorship responsibilities' for English Heritage, the National Heritage Memorial Fund, the Royal Commission on the Historical Monuments of England, the Royal Fine Art Commission and 'other heritage public bodies' (it was not stated what they were).[1] The circular continued

> To ensure that the necessary co-ordination of planning and conservation policy is maintained, formal jurisdiction on certain types of heritage casework remains with the Secretary of State for the Environment. There will be close consultation between the two Departments on many issues, including, in particular, cases raising wider policy questions, and cases where the Department of the Environment is disposed not to accept advice from English Heritage that an application should be called in, or an application or appeal rejected.

In addition to general responsibility for conservation policy, the DNH became responsible for a wide range of casework – listing, scheduling, repairs notices, reserve powers to designate conservation areas, grants to heritage bodies, responsibilities relative to ecclesiastical exemption and – a less familiar duty – responsibility for the protection of wrecks and for nautical archaeology generally. Remaining with the DOE were decisions on call-in and appeals relating to listed building and conservation area consent applications, all related enforcement, modification, revocation, purchase notice and compensation procedures under the Planning Act 1990, Article 4 Directions and proposals for demolition of churches under the Pastoral Measure. (Building preservation orders seem to have been forgotten.) Just how this judgement of Solomon was going to 'ensure that the necessary co-ordination of planning and conservation policy is maintained' was not explained. What it meant was that the responsibilities for planning and conservation which had been brought together on the formation of the DOE in 1970 were now once again split between two Departments as in the days of the Ministry of Housing and Local Government and the Ministry of Public Buildings and Works, although the separation was now more complete since the DNH acquired responsibilities for listing, conservation areas and a range of other functions under the Planning Acts which had never been part of the MPBW. In short, it divorced planning and conservation in a way that had never been done before since the Planning Act of 1909.

English Heritage Reviewed

The new chairman of English Heritage contributed a foreword to the annual report for 1991–92, although (as he acknowledged) it related to the last year of Lord Montagu's chairmanship. In it he declared that 'There is much to be done. There are more threats now to the country's heritage than ever before.' Those 'threats' can hardly have related to listed buildings, which by now were surrounded by statutory safeguards. It seems that he had in mind 'environmental problems such as acid rain and traffic pollution', although those were not the responsibility of English Heritage. The chairman continued 'I am engaged with my fellow Commissioners on a total review of all our current strategies concerned with the care of the properties with which we have been entrusted and the conservation of the fabric of England's great inheritance of buildings.' The results of that review were published in October 1992.

National Audit Office Report

Before that publication, and in time to feed into the review, the National Audit Office produced its report on *Protecting and Managing England's Heritage Property*.[2] The report provided more detailed information about the work of English Heritage than had previously been available from the annual reports. (At the time of the NAO's examination English Heritage and the other bodies concerned were the responsibility of the DOE but by the time the report was published the DNH had taken over.) The examination was not limited to the management of those properties owned by English Heritage but also dealt with their wider responsibilities for listing, scheduling and grants. The main conclusions highlighted in the report were the need to improve and computerize the lists, to make much more rapid progress on scheduling, to establish priorities for action on buildings at risk, to improve public access to buildings that received grant, and to improve presentation and visitor facilities at the properties owned by English Heritage.

The Public Accounts Committee (PAC) held hearings on the NAO report in the autumn of 1992, and its report was published in March 1993.[3] The Committee in hearing oral evidence from the Permanent Secretary of the DNH, Mr Hayden Phillips, and the Chief Executive of English Heritage, Miss Jennifer Page, gave the witnesses a fairly rough ride. It was particularly critical of the slow rate of scheduling and in setting up computerized records. It was very concerned at the results of English Heritage's preliminary survey of buildings at risk which had shown that the 37,000 buildings judged to be 'at risk and 73,000 vulnerable to neglect' included 2,400 in Grades I and II* (about 6 per cent of the total listed buildings were in those categories).[4] It also found that English Heritage had no very clear criteria for prioritizing and programming the necessary repairs to those properties (most of which were in private ownership but potentially eligible for grant assistance). There was also a backlog of £56 million repairs needed on about sixty of the historic buildings owned by English Heritage. The Committee was not satisfied with the general standards of presentation and visitor facilities at English Heritage properties, and found that the conditions were 'particularly bad' at Stonehenge, where they could only be described as a 'national disgrace'.

In general the Committee's examination was curiously narrow in focus. It hardly touched on the wider issues of conservation – what we preserve and why – such as those that the Environment Select Committee had circumvented in 1986. It touched very briefly (only two direct questions were asked) on the criteria for listing and the scope of conservation. Mr Shersby asked 'Are we as a country trying to do too much in listing too many buildings and taking on more than we can possibly cope with?' To which Mr Hayden Phillips replied that the rate of listing would certainly slow down (now that the resurvey was completed) but that some new types of buildings (such as post-war buildings) would need to be considered, and that proposals for spot listing were running at 2,500 to 2,700 a year. On the basis of this very brief exchange the Committee concluded that 'Since there is a risk that overburdening the lists with large numbers of properties selected on too broad a set of criteria may divert attention and available resources from more important properties, it would clearly be sensible for the criteria for selection and listing to be kept under careful review.'

The Committee (or some members of it) seemed to be under the impression that all listed buildings were eligible for grant and that English Heritage was in some way responsible for preserving the entire stock of listed buildings. That misconception may have been due in part to the way in which English Heritage itself had portrayed its role over the first eight years of its existence.

The government's response to the PAC's report came in the usual form of a Treasury Minute.[5] This duly gave assurance of increased effort and progress on most of the matters that had concerned the Committee. On the question of criteria for listing and the scope of the lists, the response was that 'the Department and English Heritage agree that listing cannot continue at the rate of the past 20 years, and that there is need for greater selectivity.' A new general statement of the criteria for listing would be published for consultation in the forthcoming Planning Policy Guidance Note on historic buildings and conservation, and new guidelines for assessing post-war educational buildings had recently been published, shortly to be followed by revised guidelines for listing nineteenth century buildings.

Unfortunately, only a week or so before the Committee took oral evidence, English Heritage had published the most important policy statement in its short history (see below). As this had come some three months after the NAO report, the PAC was not able (or did not choose) to focus its attention on it. What it did do was to criticize most severely the fact that English Heritage had published this key strategy document without any prior consultation with the local authorities or any of the many bodies concerned with conservation and the heritage. The chairman of the Committee said that this was 'an astonishing omission' and that he found it 'unbelievable that you did not seek the advice of all those bodies.' The explanation given was that it was thought best to present the main strategic proposals first and then to consult on their implications and practical implementation. In its report the Committee noted the explanation and reported that 'We expect them to be in full consultation with such bodies on the content and direction of the strategy before final decisions are taken, on individual properties and more generally.'

A New Strategy

The document whose lack of consultation caused the Public Accounts Committee such incredulity and concern was published in October 1992, entitled *Managing England's Heritage: Setting our Priorities in the 1990's.*[6]

This far reaching policy statement was clearly the result of the new Chairman's first six months of office. The manner of its publication bore the stamp of his impatient and assertive personality, but it also brought to bear a forceful clarity of thought and sense of strategic priorities that had been missing in conservation policy for many years – if, indeed, it had ever existed. In view of its radical approach, it was a well calculated risk to publish it and allow it to have its full impact, rather than precede it with a protracted process of consultation. The method adopted upset a lot of people and caused predictable misunderstanding, just as Mr Heseltine's proposals of 1981 had done (see chapter 18). And, just as in 1981, a lot of explaining had to be done afterwards. But as a means of arresting attention and focusing debate it had much to commend it.

The document began by announcing that the Commission had 'chosen to reassess objectives and priorities.' It continued 'To understand our past helps us to come to terms with the present and provides the foundations for the future. Our heritage plays an important educational role, but, even more importantly, a vital social role.' It went on to declare that 'the widespread concern for our heritage is to be admired and encouraged not only for its own sake, but in the interests of the social cohesion of the nation.' This was a new theme for English Heritage, reflecting the growing sociological interest in the 'heritage' phenomenon. The familiar emphasis on the economic value and tourism potential of the heritage came second. The political perception of the heritage embraced both these elements, and the notion that it had something to do with 'social cohesion' enhanced its political potential.

English Heritage was now in receipt of £100m annual grant from the government. But it had been cut by £2.6m as part of the general need for restraint in public expenditure. This no doubt served to concentrate the Commission's mind, but whereas in previous years it had pressed continuously for more funds, the opportunity was now taken to reassess the Commission's role – and not before time. Others, including the Select Committee and the PAC, might have asked whether it was sensible or practical for the Commission constantly to extend its role and its expenditure to match the ever extending scope of the heritage. To its credit, the Commission seems to have asked this question of itself.

In a section headed 'Making the Heritage Pound go further', which combined a messianic note with an awareness of practicality, the Commission announced that:

> We will focus our powers, our skills and our resources where the need is greatest; where the most important historical and architectural buildings and sites are at risk. We will pay particular attention to the places where we live and work, which affect the people we are and can become.
>
> To be most effective we will step back from work which, although important, is better done by others and we will work increasingly as an enabler at local level.

This notion of 'stepping back' from some areas of its traditional work caused immediate consternation to the heritage lobby. But worse was to come as the Commission explained what it had in mind:

> We will divide our historic properties and sites into three categories of importance and concentrate on the first two categories. We will reduce expenditure on properties and sites in the third category and seek to pass these over to the local management wherever possible.

The Commission then went on to announce that it would focus its resources on Grade I and II* buildings (about 6 per cent of all listed buildings), and make conservation area grants 'only in areas which combine townscape quality with financial, material and social need.' Churches would get grants only where they could raise complementary private funding. Funds would be diverted from rescue archaeology to a small selection of longer-term projects. The Commission's statutory work (inherited from the GLC) on Grade II buildings in London would be transferred to the Boroughs (in line with arrangements outside London). Efforts would be made to set up a Conservation Fund with 'major private foundations and others' to harness private resources to enable English Heritage to take emergency action to save outstanding properties where other agencies were unable to help. The conservation of Stonehenge would be ensured using private sector resources. The Commission would work with other 'interested bodies' to raise special funds to preserve 'significant historic entities', and would launch a public appeal for funds to set up a permanent grant scheme for historic gardens.

To conclude this staggering manifesto the Commission produced a quotation that must have caused almost as much pain to many as the apparent intention to privatize the heritage:

> Our Fathers in a wondrous age
> Ere yet the Earth was small
> Ensured to us an heritage
> And doubted not at all,
> That we the children of their heart,
> Which then did beat so high,
> In later times should play their part
> For our posterity.
>
> Rudyard Kipling

The Response

The response to these proposals from the conservation bodies and others concerned with the heritage was at least as violent as that which had greeted the original proposals that led to the setting up of English Heritage (see chapter 18).

On 18 May 1993 the Commission published a set of four press releases and the text of the Chairman's speech to the Association of District Councils' Heritage Conference. This was its response to the furore caused by the strategy document. It sought to repair some of the damage caused and to put flesh on the bones of the stark policy initiatives set out in the earlier publication.

Mr Steven's speech to the conference was headlined 'Planning a Future for England's Heritage'. He began by referring to the criticisms, often highly personalized, which last October's bombshell had provoked: 'All of you in political life know very well the poisoned barbs that are chucked at you from time to time.' But he quickly moved on to his main theme 'Prioritizing the Heritage'.

The Commission was 'shifting our approach from where we are involved quite unnecessarily in minor detail and turning our attention to a more deliberate strategic approach – to the high ground.' He then went on to explain the proposals for the devolution of site management, changes in the conservation area grant regime, the devolution of powers in London, and the future role of English Heritage.

On the widely misunderstood subject of devolved site management, he said 'we are definitely not "selling off the silver plate".' The idea was simply to find local authorities or other suitable bodies to take on responsibility for some properties and sites that might be better managed locally than by English Heritage. This was not a radical change but a policy of some years' standing. Some twenty properties had already passed to local management in recent years and seventeen properties were jointly managed with the National Trust. Already, since publication of the strategy, there had been expressions of interest in eighty-four properties from local authorities and other bodies. English Heritage would retain final responsibility, total expenditure would not be reduced, public access would be maintained and detailed local agreements would be drawn up in each case. Nothing to get too excited about here.

The devolution of listed building control for alterations to Grade II buildings would bring London into line with the rest of the country, and would depend on English Heritage being satisfied that each Borough was professionally equipped to do the job.

On conservation areas the proposals were spelt out in detail in a consultation paper: *Conservation Area Partnership Scheme.*[7] In 1992–93 English Heritage spent £7.3 million on conservation area grants. But the variety of grant schemes was confusing and the grants were too dispersed or 'pepper potted'. In future the main channel for grants would be new Conservation Area Partnerships, which would be based on 'specific programmes of identified work', and with more explicit criteria for the allocation of funds, measurable targets and timescales. There were at present nearly 400 schemes of various types operated with local authorities, and the aim would be to convert the best of these to the new Partnership model. This was all very reasonable but then a sternly realistic note was struck: 'English Heritage has the capacity and the resources, however, to become involved only in a small fraction of the country's estimated 7,500 conservation areas.'

In short, what all this meant was that English Heritage was no longer to be regarded as, and would no longer present itself as, the all-powerful, all-providing patron of the heritage. Since concern for the heritage was now so widespread, and as so many people had a vested interest in their local heritage, local people and organizations must accept more responsibility for it. The Chairman did not quite say as much (or not in public) but this is what it meant. It was an overdue douche of realism for both English Heritage and its clients.

National Heritage Committee

Two months after English Heritage had published its new strategy document, the House of Commons National Heritage Committee announced its intention to undertake an inquiry into English Heritage, and that it intended to concentrate on the policy document. For one reason and another, not least the concurrent inquiry by the Public Accounts Committee, the National Heritage Committee's inquiry did not get underway until over a year later. The Committee then announced that it intended to extend its inquiry to cover English Heritage's contribution to tourism and, in particular, to look at maritime heritage. Some 140 memoranda were submitted to the Committee, and oral evidence was heard on six days in January to February 1994. The Committee's report was published on 3 March,[8] together with the minutes and a selection of written evidence that ran to 250 pages.

The National Heritage Committee's inquiry was the least enlightening of the three Parliamentary inquiries concerning English Heritage that had taken place over the past three years. The Committee seemed to have difficulty focusing its attention on any one of the diverse topics that interested it. In conducting its hearings it professed puzzlement about how responsibilities for the heritage were divided as between the Department of National Heritage, English Heritage and the various other organizations, museums and voluntary bodies that seemed to be involved with the heritage. It elicited the fact that the DNH had taken over responsibility for forty-eight agencies and non-governmental organizations from six different Departments. The Committee's confusion was perhaps understandable since, in these early years of 'agencization', the transition from the traditional Whitehall way of doing business to a more structured system of relationships and accountability had not yet been thought through (and is still far from clear).

Much of the written material submitted to the Committee duplicated what had already been examined by the Environment Select Committee and the PAC. Not a great deal had happened in the interim. The Committee's hearings of oral evidence certainly generated more heat than light, particularly when the Chairman of English Heritage, Mr Jocelyn Stevens took the stand. Some members of the Committee indulged in a little bear-baiting, implying that the Chairman was not fighting English Heritage's corner hard enough. This provoked Mr Stevens to a memorable outburst:

> I am obsessed by the battle to save our heritage. Everything is a fight, a fight against planners, the experts; Government, one Department not agreeing with another Department, all that . . . I fight from dawn to dusk. The Chief Executive is in the next door office to mine and I scream and shout all day and she has now caught it, we both scream and shout all day. It is not a soft bed to lie in.

Most of the Committee's questioning, and the evidence that witnesses wished to give, resolved into questions of funding: how much should be spent on the heritage and by whom. But these questions, of course, raised other questions – what is meant by 'the heritage', what does it include, why do we want to save it, whose responsibility is it? The Chairman of the Committee said to the Chairman of English Heritage 'your corner is part of an utterly uncoordinated and unthought-out whole.' To which Mr Stevens replied 'Thank you Chairman. I think that was a very helpful statement, if I may say so.'

One member of the Committee, Mr Maxton, tried to raise some of the policy issues when he put it to Mr Stevens that 'It seems to me that you have no overall strategy in your organisation as to why you preserve things, what things you preserve.' He suggested that there ought to be an 'overall plan in terms of what you ought to be preserving in this country, what is worth preserving and saying that on occasions some things are not worth preserving and they have to go.' But without waiting for any response to these questions, Mr Maxton moved on directly to 'the whole question of tourism', and the Committee did not return to the larger policy issues.

Mr Stevens produced two bits of statistical evidence that may have strained credibility. A Gallop poll carried out in August showed that 40 per cent of the population had visited a heritage site within the last three months, and that this was double the number that did so in 1989. The poll also revealed that over half the population felt that the government and local authorities should provide 'far more money to help preserve historically interesting buildings and sites' (perhaps more interesting was that 27 per cent disagreed).

But Mr Stevens could not help the Committee in assessing how much needed to be spent on the heritage. Committee member Mr Sykes put it to him 'Is it not the case that really the heritage could be a bottomless pit for British taxpayers' money, and they could never ever have enough money for all you really want to do?' To which Mr Stevens replied 'I could not but agree with you.' He went on to say 'What I am pleased about is – I think we are developing a political realisation . . . Heritage should move up the political agenda – it deserves more support. If we do not get more support then it will disappear and it is too late then to recreate it.'

At the end of its inquiry the Committee reached no clear conclusions on what was meant by 'the heritage', how much needed to be spent on it, by whom, for what purpose, or how much the nation could afford. It offered a few recommendations on some of the more detailed matters that had been raised (such as VAT on repairs and maintenance, craft training and maritime archaeology), but its recommendations on the wider issues only reflected the Committee's inability to discover what current policies and priorities, if any, existed:

> The Committee recommends that consistent criteria be applied at Department of National Heritage level so that some undertakings are not funded more or less generously than others.
>
> The Committee calls upon the Secretary of State in his reply to this report to define and focus more adequately the department's policy on national heritage.
>
> The Committee believes that the Department of National Heritage should now publish a clear policy on marketing, funding and, above all, on priorities.

The Government's Response

In its response to the National Heritage Select Committee, the government expressed some irritation that the Committee had failed to recognize that 'substantial progress' had been made, in the two years since the DNH had been created, 'in the articulation of coherent policies for the protection of the national heritage.'[10] Those policies, it was claimed, had been set out in the speech by the Secretary of State given at the Royal Fine Art Commission on 3 December 1992, which had since been published.[11] They would be 'restated' in Planning Policy Guidance 15 which was to be published later in the year (see chapter 22).

The government also observed that the Committee's report 'pays curiously little attention to the likely impact of the National Lottery', which was forecast to yield on average more than £200 million a year for the heritage. The task of distributing this huge sum (nearly double English Heritage's annual budget) has since been entrusted to the National Heritage Memorial Fund, which was itself established in 1980.[12] The NHMF will administer its Lottery Fund separately from the modest £12 million a year which it has had available in previous years. In 1994 the NHMF published guidelines for potential applicants in which it explained that

> The primary aims of the Heritage Lottery Fund are to secure, conserve and improve assets of importance to the national heritage, whether land, buildings or objects, and to enhance public access to and enjoyment of such assets. It will consider projects falling within the whole range of its remit, including ancient monuments, historic buildings and their contents and settings, designed landscape, land of scenic or scientific importance, special library collections and industrial, transport and maritime history.

Thus the NHMF emerges as by far the wealthiest patron of the heritage, and its potential customers are legion. Whether it makes sense to divert so much money to a body with no wider policy responsibilities for the heritage and with an eccentric record of patronage is open to doubt. In 1995 it at once attracted widespread criticism by announcing that as its first grant from the Lottery Fund it had paid £12.5 millions to the family of Sir Winston Churchill for a collection of his personal papers (many of which were replicated in the government's own records). The Chairman of the Fund, Lord Rothschild, commented wearily that 'We are in a no-win position. On nearly every aspect of heritage there are at least two points of view.'[13] No doubt the National Heritage Select Committee will be returning before long to see what effect the Lottery bonanza is having on heritage policy.

NOTES

1. Department of the Environment (1992) *Responsibilities for Conservation and Casework*. Circular 20/92, Department of National Heritage, Circular 1/92. London: HMSO.
2. National Audit Office (1992) *Protecting and Managing England's Heritage Property*. Report by the Comptroller and Auditor General. London: HMSO.
3. Committee of Public Accounts (1992) *Protecting and Managing England's Heritage Property*. 19th Report, House of Commons, Session 1992–93. London: HMSO.
4. English Heritage (12992) *Buildings at Risk: A Sample Survey*. London: English Heritage.
5. Treasury Minute on the Twenty-ninth to Thirty-first Reports from the Committee of Public Accounts, Command 2257. London: HMSO.
6. English Heritage (1992) *Managing England's Heritage: Setting Our Priorities for the 1990's*. London: English Heritage.
7. English Heritage (1993) *Conservation Area Partnership Scheme: A Consultation Paper*. London: English Heritage.
8. House of Commons, Session 1993–94, National Heritage Committee, Third Report (1994) *Our Heritage: Preserving It, Prospering from It*, Vol. I, *Report and Proceedings* [HC 139]. London: HMSO.
9. House of Commons, Session 1993–94, National Heritage Committee, Third Report (1994) *Our Heritage: Preserving It, Prospering from It*, Vol. II, *Minutes of Evidence and Appendices* [HC 139–II]. London: HMSO.
10. House of Commons, Session 1993–94, National Heritage Committee, First Special Report (1994) *Government Response to the Third Report of the National Heritage Committee* [HC 549]. London HMSO.
11. Brooke, Rt. Hon. Peter (1992) *Shaping Our Heritage*. London: Department of National Heritage.
12. National Heritage Act 1992, 2 Eliz. ch. 17.
13. Lord Rothschild, Interview in *The Independent*, 10 May 1995.

22

ADVICE AND DOCTRINE

We have already seen how the Ministry of Health and its successors allowed the enactment of successive Planning Acts from 1909 to 1932 to pass without offering any advice or guidance to local authorities on the provisions relating to buildings of special architectural or historic interest. With the post-war Acts, brief explanatory notes were issued but it was not until the first Planning Bulletins were published in 1962–63 that the MHLG had anything to say on the subject; and then only briefly and in the context of town centre redevelopment (see chapter 11). This habit of restraint in official pronouncements was to change dramatically over the next twenty years, when circulars proliferated and the thin stream of advice on conservation became a flood.

Much of the advice issued in those years was directed at raising the general awareness, especially among local planning authorities, of the importance and potential scope of conservation. It was generally uncontentious and, as circular followed circular, began to suggest a tendency to play to the heritage gallery. This was what those concerned for conservation wanted to hear, and successive governments showed no inhibition in preaching to the converted. Sound advice bears repetition and it is notable that large swathes of policy guidance were lifted and repeated from one circular to another, generally verbatim.

In terms of policy development, the interest is not in the continuum of conventional observations on the need for conservation but in the occasional change of emphasis and in the emergence of certain points of doctrine or tenets of conservation policy.

TWENTY YEARS OF CIRCULARS: 53/67 TO 8/87

The first circulars of substance were 53/67 on the Civic Amenities Act 1967 (which introduced conservation areas) and 61/68 on Part V of the Town and Country Planning Act 1968. Both of these circulars have already been referred to in chapters 13 and 14. They began a tradition of prolixity that was fully developed in later years. Circular 53/67 made a modest start with thirty-six paragraphs and fifteen more in an annex: a total of about 4,500 words. Circular 61/68 ran to nearly 10,000 words. Ten years later Circular 8/87 notched up about 27,500 words, and the latest compilation, Planning Policy Guidance 15 of 1994, totals around 34,000 words. These are rough estimates, but they make the point.

As noted in chapter 13, Circular 61/68 introduced what became known as 'the presumption in favour of preservation'. It did so in terms that conjured up the powerful spectre of a country devoid of historic buildings:

> Generally it should be remembered that the number of buildings of special architectural and historic interest is limited; that the number has already been reduced by demolition, especially since 1945; and that unless this process is, if not completely arrested, at least slowed down, the virtually complete disappearance of all listed buildings can be predicted within a calculable time. Accordingly the presumption should be in favour of preservation except where a strong case can be made out for a grant of consent after application of the criteria mentioned.

This domesday scenario was patently unrealistic. The listing system itself erected barriers against unchecked demolition, and the number of listed buildings was being steadily increased. Many (probably most) listed buildings maintained their market value, and the cachet of listing tended to enhance it, especially for residential property. It seems most unlikely that the annual rate of historic buildings demolished had increased since 1945, when one considers how few buildings pre-1700 remained by 1940 and how many eighteenth- and nineteenth-century buildings were destroyed by war damage. Clearly any accelerating rate of erosion leading to 'complete disappearance of all listed buildings within a calculable time' was a figment of the imagination. It was the kind of tendentious rhetoric associated with pressure-group propaganda rather than with government circulars. But it fuelled the belief that the loss of any listed building would contribute to the rapid extinction of an endangered species. The conservationist lobby seized on this doctrine, which the government had so conveniently articulated, and it was to be repeated (with minor variations) in all subsequent policy guidance on the subject.

Circular 86/72 extended the doctrine much more widely: 'in the same way, the presumption should normally be in favour of demolition control in a Conservation Area.' As recorded in chapter 14, Circular 46/73 went even further and urged that conservation policies should 'take account of the growth of public opinion in favour of conserving the familiar and cherished local scene. It should also have care for the conservation of existing communities and the social fabric.' Later circulars, however, did not repeat this particular effusion of conservationist sentiment or, at least, couched it in somewhat less emotive terms.

Circular 102/74 followed on the reorganization of local government and was aimed at instructing the new and enlarged district planning authorities on their responsibilities for conservation. It began by acknowledging that 'the development and conservation policies of many of the former local authorities reflected the growing public concern that the national architectural heritage was being rapidly eroded.' In tune with the tone of Circular 86/72 it emphasized that 'historic buildings once lost are irretrievable.' That, of course, is a truism but it fostered the assumption that any building once listed ought to be preserved. The circular went on to try to counter that assumption: 'It is perhaps necessary to emphasise again that listing does not itself imply that the building will necessarily be preserved, but merely requires that the building's case for preservation is examined under the statutory procedures for listed building consent applications.' But, while that was the statutory position, the policy was clearly biased strongly in favour of preservation.

The policy advice and doctrine built up over the previous twenty years was consolidated in the omnibus Circular 8/87 on *Historic Buildings and Conservation Areas: Policy and Procedures*. This was to be the conservation bible for the next seven years.

The circular began with the assurance that 'there are essentially no changes in the Government's policy', and that 'To ensure continuity of policy advice, the wording of the earlier circulars has been retained wherever appropriate to do so.' That was broadly true but, while the sacred texts were largely preserved (and the minor amendments did not appear specially significant), a great deal more was added. Public opinion was now said to be '*overwhelmingly* in favour of conserving *and enhancing* the familiar and cherished local scene' (amendments to Circular 46/73 italicized).

The new circular, however, did introduce a new element prompted by the government's concurrent concern for renewal and economic regeneration:

> previous circulars have stressed the need to preserve our architectural heritage in different ways, but the main message is clear; if we do not take steps to protect and preserve buildings of value, either in their own right or because of the contribution they make to a pleasant townscape or village scene, they may well be lost, and once lost, they cannot be replaced. It should, however, be remembered that as the Secretary of State has pointed out, *our heritage is the product of many centuries of evolution and it will continue to evolve. Few buildings exist now in the form in which they were originally conceived. Conservation allows for change as well as preservation.* There are many cases where it is right to 'conserve as found'. But there are circumstances too where our architectural heritage has to be able to accommodate not only changes of use but also new building nearby. It is better that old buildings are not set apart but are woven into the fabric of the living and working community. This can be done provided that the new buildings are well-designed and follow fundamental architectural principles of scale and the proper arrangement of materials and spaces and show respect for their neighbours. Conservation means breathing new life into buildings, sometimes by restoration, sometimes by sensitive development, sometimes by adaptation to a new use and always, by good management. Taking decisions on matters concerning listed buildings and conservation areas involves balancing many factors and the guidance given in this circular will, it is hoped, help authorities to form sensible judgements on questions which arise in their areas.

A far less pragmatic and distinctly doctrinaire approach was evident in other parts of the circular. For example, while elsewhere urging the re-use of old buildings, the circular pronounced that 'The best use for an historic building is obviously (*sic*) the use for which it was designed and wherever possible this original use, particularly if it is residential use, should continue. If the use of the building has been changed from its original purpose, it should be considered whether it can revert to it.' This is the voice of conservation rampant.

The bulk of the circular, however, comprised detailed and helpful advice on all aspects of listed building and conservation area procedures, which showed that the Department had now adopted an almost paternalistic attitude towards local planning authorities in advising them on their conservation work. It is also very apparent that, at a time when the government had for nearly ten years been pursuing deregulatory policies and seeking ways of simplifying the planning process, a very different attitude was taken towards conservation. The general impression created was that 'planning' was a bad thing, or at best a necessary evil, whereas 'conservation' was a good thing and to be fulsomely encouraged. This dichotomy between planning and conservation led to increasing emphasis on the latter at the expense of the former, whereas twenty years previously in Circular 53/67 Ministers had said that they attached 'particular importance' to 'creative planning for conservation'. This had been the message of Planning Bulletins 1 and 4 in 1962 and of the 1967 landmark publication *Historic Towns: Preservation and Change* (see chapter 13).

Criteria for Listing

Circular 102/74, for the first time in a Departmental circular, set out the basis on which the lists of buildings of special architectural or historic interest were compiled. As the circular noted, 'The listing standards are revised from time to time.' This was certainly true. The original system was devised by the Maclagan Committee (see chapter 8) but details of it were never published. An amended version was included in the MHLG's report for 1943–51 (see chapter 10). Circular 8/87 included a substantially revised and much fuller explanation of the 'principles of selection' and this is reproduced in Appendix C together with the earlier version from Circular 102/74 and the latest version, again substantially revised, from PPG 15 of September 1994. These successive revisions reflect the expanding scope of conservation interests as they extend to include buildings of later periods and a wider range of building types. The practice of moving and widening the goal posts in this way can be confusing. It necessitates a parallel process of continually revising the lists and prompts demands for spot-listing of individual buildings that have either been overlooked in earlier lists or are now thought to be within the scope of the latest listing criteria. That is probably inevitable if the 'principles of selection' are not to be set in concrete but are to respond to changes in architectural taste or scholarship and to changing perceptions of the heritage.

Circular 8/87 prevailed for seven years until it was replaced by PPG 15 in 1994 (see below). The only change between those dates was in Circular 18/88, which recorded a decision of the High Court that contrary to earlier advice, 'repainting was capable of being an alteration' and therefore could require listed building consent. While an Englishman's home had long since ceased to be impregnable so far as conservation procedures were concerned, it had been thought that at least he could choose the colour of his front door: that was no longer necessarily so.

Planning Policy Guidance 15

PPG 15 was published in September 1994 and is now the *vade mecum* for conservation policy.[1] As it is readily available from HMSO there is no need to paraphrase its contents in detail here. But two aspects are worth particular notice in the context of this study.

First it is notable that this comprehensive advice on conservation does not refer to conservation in its title but is *Planning Policy Guidance: Planning and the Historic Environment*. Thus it can be taken to reflect a wish to stress the essential interrelationship between planning and conservation. It was issued jointly by the Department of the Environment and the Department of National Heritage.

Secondly, it is interesting to see whether this new guidance simply re-iterates the key points of doctrine referred to earlier in this chapter; in particular, the 'presumption in favour of preservation'; the insistence that the original use of a listed building is the best use and to be continued if at all possible; and the assertion that the Secretary of State would not consent to demolition unless he was satisfied that every possible effort had been made to continue the present use or to find a suitable alternative use for the building.

These were three key components of Circular 8/87. When PPG 15 was issued in draft form for consultation in July 1993, these were touchstones against which continuity of policy and the government's commitment to conservation were likely to be judged. Large parts of earlier policy guidance were reproduced, as they had been in earlier circulars, so the omission or alteration of any of these three components could be easily identified.

The Presumption

Circular 8/87 (para. 91) said that 'the Secretary of State is of the view that the presumption should be in favour of preservation except where a strong case can be made out for granting consent after application of the criteria mentioned.' Those criteria related to the importance of the building both intrinsically and relatively (i.e. in relation to the number of other listed buildings in the area); its architectural and historic interest; the condition of the building and the cost of repairs; the importance of any alternative use for the site for a public purpose; and 'whether, in a rundown area, a limited redevelopment might bring new life and make the other listed buildings more viable.'

The draft PPG did not repeat the 'presumption in favour of preservation'. But it repeated all the earlier considerations that would weigh in favour of preservation, added some new ones and found different ways of saying the same thing in various parts of the draft. A footnote to para. 3.3 explained why the 'presumption' had been abandoned:

> paragraph 91 of Circular 8/87 expressed the policy in terms of a 'presumption in favour' of the preservation of listed buildings. For reasons explained by Sir George Young as Minister for Planning in an article entitled 'A Plague on presumptions' in the December 1991/January 1992 issue of Housing and Planning Review, that formulation is now avoided. No change is intended in the importance attached to the preservation of listed buildings.

The article by Sir George Young was not explicitly related to the 'presumption in favour of preservation' but concerned the 'presumption in favour of development', which was incorporated in Planning Policy Guidance Note No.1 of 1988, although it dated back to the debates on the 1947 Act. What it meant was simply that the developer was entitled to his planning permission unless there were good policy reasons for refusing it. The word 'presumption' had also appeared in circular 14/84, which spoke of a 'general presumption against inappropriate development within Green Belts' (this formula did not appear in the original Green Belt circulars of 42/55 and 50/57). Sir George Young acknowledged that 'presumption' (for or against) had for decades been a convenient form of expression, but suggested that 'it can be used as a substitute for thought' and could 'all too easily lessen the scope for a flexible approach to development control.'

Sir George Young's 'a plague on presumptions', however, could not prevail against the conservation lobby. In the final version of PPG 15 the 'presumption' was reinstated:

> There should be a general presumption in favour of preservation of listed buildings, except where a convincing case can be made out, against the criteria set out in this section, for alteration of demolition

The criteria in PPG 15 enlarge on those in circular 8/87, but the 'presumption' remains in place.

Original Use

Circular 8/87 (para. 20) had said that 'The best use for an historic building is obviously the use for which it was designed and wherever possible this original use, particularly if it is residential, should continue.'

That sweeping statement was not reproduced in the new PPG. Both the draft and final version took a more realistic and pragmatic line:

> 3.10 The best use will very often be the use for which the building was originally designed, and the continuation or reinstatement of that use should certainly be the first option when the future of a building is considered. But not all original uses will now be viable or even necessarily appropriate: the nature of uses can change over time, so that in some cases the original use may now be less compatible with the building than an alternative. For example, some business or light industrial uses may now require less damaging alterations to historic farm buildings than some types of modern agricultural operation. Policies for development and listed building controls should recognise the need for flexibility where new uses have to be considered to secure a building's survival.

Alternative Use

A related change was that whereas Circular 8/87 (para. 89) said that the Secretary of State would not be prepared to grant listed building consent for the total or substantial demolition of 'a' (*sic*) listed building unless he was satisfied that every possible effort has been made to continue the present use or to find a suitable alternative use for the building, the new draft PPG changed 'a' to 'any'. The word 'any' was retained in the final version of PPG 15. While this might seem to make no difference grammatically, the draft printed 'any' in bold type and underlined – the intention presumably being to reinforce this particular tenet of conservation policy doctrine.

Deregulation

While Circular 8/87 had moderated some of its more extreme statements of conservation policy by noting the need for some flexibility as regards alterations and change of use where that would facilitate conservation, the new PPG attempted to reflect something of the government's other policies on deregulation and simplification of the planning system and the need to promote economic regeneration and renewal. These potentially conflicting objectives were not successfully reconciled in the new guidance. The drafting tended to swing rather violently between doctrinaire conservation and the contrary concern for economic development. This uneasy contrast was more apparent in the draft PPG than in the final version, where attempts had evidently been made to blur the distinction and redress the balance. For example, the draft PPG opening paragraphs read as follows:

The function of the planning system is to regulate the development and use of land in the public interest. It has to take account of the Government's objective of promoting economic growth, and make provision for development to meet the economic and social needs of the community. As PPG1 makes clear, planning is also an important instrument for protecting and enhancing the environment in town and country, and preserving the built and natural heritage. It is essential that planning processes take full account of both these objectives – the need for economic growth and the need to protect historic buildings and areas, which are part of our sense of national identity and are valuable for their own sake and for their role in education, leisure and tourism.

These two objectives should not be seen as necessarily in opposition. Conservation of our historic and natural environment can play a key part in promoting economic prosperity, for instance by fostering the growth of tourism and by providing the attractive living and working conditions which encourage inward investment in an area. Environmental quality is increasingly a key factor in many economic decisions.

In its final version PPG 15 had a new opening paragraph that banged the drum for conservation before moving on to a slightly modified version of the earlier draft:

1.1 It is fundamental to the Government's policies for environmental stewardship that the there should be effective protection for all aspects of the historic environment. The physical survivals of our past are to be valued and protected for their own sake, as a central part of our cultural heritage and our sense of national identity. They are an irreplaceable record which contributes, through formal education and in many other ways, to our understanding of both the present and the past. Their presence adds to the quality of our lives, by enhancing the familiar and cherished local scene and sustaining the sense of local distinctiveness which is so important an aspect of the character and appearance of our towns, villages and countryside. The historic environment is also of immense importance for leisure and recreation.

1.2 The function of the planning system is to regulate the development and use of land in the public interest. It has to take account of the Government's objective of promoting sustainable economic growth, and make provision for development to meet the economic and social needs of the community. As PPG1 makes clear, planning is also an important instrument for protecting and enhancing the environment in town and country, and preserving the built and natural heritage. The objective of planning processes should be to reconcile the need for economic growth with the need to protect the natural and historic environment.

Evidently the conservation lobby had done its work on the draft document and had proved more persuasive than those speaking for economic development and renewal. The balance in the final version of PPG 15 was tipped heavily in the direction of conservation.

The attempt to square the circle (i.e. 'to reconcile the need . . .') by evoking the contribution conservation can make to tourism is disingenuous. In fact the impact of tourism can have a drastic effect on attempts to preserve the character of an historic area (as it can on places of natural beauty). This is a well known dilemma and not one that is easily resolved.

These criticisms of the drafting of PPG 15 may seem pedantic, but such ambiguities, as well as doctrinaire policy statements, are meat and drink to the legal profession that feeds off the planning system. The truth is that in this area of planning policy, perhaps more than most others, attempts at generalization – especially when expanded at great length and in various versions – are not always very enlightening and often provide fertile ground for sterile controversy and legal obfuscation. With conservation, as with many other aspects of planning, what matters is not the application of general and imprecise principles but the careful analysis and assessment of particular proposals in the context of the local environment and in relation to local objectives, which ought to be reflected in the local development plan. The merit of the British planning system is that it allows for this and the process is not really helped by the fashion in recent years of delivering ever-more verbose policy guidance from the Departmental oracle.

Design

PPG 15 contains a wealth of sound advice on conservation but it retains elements of conservation doctrine that can impede rather than promote effective planning and conservation practice.

One feature introduced into PPG 15 struck a new note that must be very unwelcome to those who are concerned as much with the cultivation of good design in new building as with the protection of the best in the historic environment. The new guidance advised that one of the 'general considerations' to be weighed in assessing a proposal for demolition of a listed building was 'the merits of alternative proposals for the site'. But the advice given was hardly calculated to encourage creative architecture:

While these are a material consideration, the Secretaries of State take the view that subjective claims for the architectural merits of proposed replacement buildings should not in themselves be held to justify the demolition of any (*sic*) listed building.

This makes depressing reading and one notes the disparaging reference to 'subjective claims for the architectural merits of proposed replacement buildings'. This is as if to say that the merits of listed buildings are in no way a matter of 'subjective judgement', or that the appraisal of new architecture is necessarily less informed by knowledge, expertise and discriminating judgement than the assessment of the value or 'special interest' of older buildings.

This deplorable new doctrine may well have been the conservationists' revenge for their defeat in the case known variously as No 1 Poultry, or the Mansion House Square scheme or the Mappin and Webb site. This was (and remains) a classic case in the history of planning and conservation and needs to be described in some detail.

No 1 Poultry

This case, which is of exceptional interest and importance, lasted seven years. It began as a planning issue concerned with aesthetic (design) control and ended as one concerned with conservation.

The site lies at the heart of the City of London, at the junction of seven roads, adjacent to the Mansion House, the Bank of England and other prominent historic buildings, and less than half a mile east of St Paul's Cathedral. The site itself is a little less than one acre in extent, with Poultry and Cheapside to the north and Queen Victoria Street on the south east. The site contained eight listed buildings (all Grade II) and several other buildings not listed but all within a conservation area. The listed buildings were late nineteenth-century commercial buildings and offices, typical of the period. At the time when they were built *The Builder* referred to them scornfully as a debased example of Ruskinian gothic, with an excess of mass-produced decorative stonework. But they had become a city landmark, partly because of their prominent position on a triangular site and partly because the front portion had for many years been the premises of Mappin and Webb, whose windows full of silver-plate added a gleam of mayoral glitter to an otherwise dingy corner of the city.

For many years a property developer, Mr Peter (now Lord) Palumbo had been assembling the site for redevelopment. His father had also been a successful property developer and Mr Palumbo had the ambition of providing London with a building of outstanding architectural quality. He had commissioned the most distinguished surviving representative of the Modernist Movement, Mies Van der Rohe, to design a building for the site. The result was a design for a thirty-storey skyscraper, strongly reminiscent of Van der Rohe's buildings on Chicago lakeside. Some opponents claimed that it had been designed for that purpose or that the aged architect had adapted an earlier design for the London site. No less an architectural critic than the Prince of Wales, in his notorious speech to the RIBA, described it as a 'glass stump' more suited to Chicago than the City of London. Be that as it may, it was a design of great distinction and refinement, unlike anything then existing in central London (the NatWest tower nearby was mediocre by comparison).

The City of London, as planning authority refused planning permission, primarily on grounds of its being grossly out-of-scale with its surroundings, incongruous and overbearing. The case then went to the Secretary of State on appeal and to a public inquiry in 1984. While agreeing that the proposed design was unsuitable for the site, in his decision letter the Secretary of State said 'it would be wrong to attempt to freeze the character of the City of London for all time'; that he did not 'rule out development of this site if there were acceptable proposals for replacing the existing buildings'; and that he did not 'consider that the buildings are of such overriding importance that their preservation should outweigh all other considerations.'

Acting on the Secretary of State's strong indication that a redevelopment of less dominant character might be acceptable, Mr Palumbo next commissioned a new design from the famous British architect James Stirling. His design was more modest in scale and in the somewhat eccentric post-modernist style of which Stirling was a leading exponent; he also emphasized its 'contextual' character.

The new design was generally applauded by the architectural establishment and the Royal Fine Art Commission. The Prince of Wales again contributed his opinion, dismissing the proposed building as being like a 1930s radio. But this time the opposition was based not primarily on aesthetic grounds but on conservation and the special architectural interest of

No 1 Poultry, City of London (1871): demolished for redevelopment as offices (1995).

the listed buildings that would be demolished.

The new scheme was subjected to as protracted a series of administrative and legal procedures as any development project can encounter. First the City of London refused planning permission and the developers' appeal was heard at a public inquiry which started on 17 May 1988. The arguments turned chiefly on whether the listed buildings were of such importance as to outweigh the case for redevelopment, and whether the proposed new building was of sufficient quality to override the case for conservation. The objectors were, of course, able to cite the two key points of policy guidance from Circular 8/87, already highlighted in this chapter. First the 'presumption in favour of preservation', and secondly that 'The Secretary of State will not be prepared to grant listed-building consent . . . unless he is satisfied that every possible effort has been made to continue the present use or to find a suitable alternative use for the building.'

Among the objectors, both English Heritage and SAVE (the conservation group Save Britain's Heritage) sought to show that a viable scheme for reconstructing the listed buildings within the existing facades was feasible, and that the possibility of retaining the buildings in use had not been proved to be impractical. The developers eventually conceded that retention was possible but sought to establish that the proposed scheme was of such quality that it would enhance the conservation area.

The Inspector who held the inquiry was himself a professional architect and town planner. The public inquiry lasted for eighteen working days and the Inspector's report ran to 131 closely typed pages.[3] The following excerpts from his concluding paragraphs indicate how clearly he focused on the issues at stake:

> I come now to consideration of the criterion on which I believe the appeal proposals either stand or fall – whether or not they are of such greater intrinsic quality than what could be achieved by the refurbishment or reconstruction of the existing buildings on the site that it is important that they should be permitted.
>
> Taken overall, as far as the public domain is concerned, I would say that the design has strong consistency and character and is one which would be a worthy modern addition to the architectural fabric of the City. It might just be a master piece . . .
>
> A new scheme must have outstanding qualities if it is to overcome the strong presumption in favour of the retention of listed buildings and the attractive opportunities that conservation can offer . . .
>
> It is my assessment that the appeal proposals, by their dignified order, their imaginative ingenuity and pervading overall consistency, would contribute more both to the immediate environment and to the architectural heritage than the retention of the existing buildings . . .
>
> The loss (of the opportunity to build Stirling's design) would be all the greater because it would be of a considered mature work by a British architect of international stature of whose achievements the nation can justly be proud. In my view it deserves to be built.

The Secretary of State in his decision letter following the appeal quoted the Inspector's conclusions in full.[4] In a very condensed summary of the issues involved (which was later to be criticized) he indicated that he agreed with the Inspector and concluded that the proposed redevelopment scheme

> would contribute more, both to the immediate environment and to the architectural heritage, than would the retention of the existing buildings, and that in view of this he is satisified in the special circumstances of this case that the usual presumption, in favour of preservation of listed buildings, should be overridden.

He therefore allowed the appeal. But this was only the start of the saga. SAVE appealed to the Crown Court against the Secretary of State's decision, but the Court upheld the decision.[5] The case then went to the Court of Appeal who reversed the lower Court's decision.[6] Finally the developers appealed against that decision to the House of Lords.[7] It is not necessary to repeat all the arguments rehearsed by the various parties before the Crown Court and the Court of Appeal. The relevant issues were fully considered in the judgement of Lord Bridges in the House of Lords. The precise issue on which their Lordships were asked to decide was (as Lord Ackner pointed out) a narrow one – viz, not whether the new building was sufficiently meritorious to override the presumption in favour of preservation (this aesthetic judgement was for the Secretary of State in deciding the planning appeal), but whether the Secretary of State had complied with rule 17(1) of the Town and Country Planning (Inquiries Procedure) Rules 1988 by providing adequate reasons for his decision. In brief, their Lordships decided that the decision should not be overturned on those grounds and they reversed the Appeal Court's decision. But in delivering his judgement Lord Bridges reviewed the long history of the case and considered all the wider issues that had been raised.

As Lord Bridges said 'At the heart of the policy issue were passages from Circular No 8/87', and he quoted the relevant part of paragraphs 89 ('unless he is satisfied that every possible effort has been made to continue the present use' etc.); para. 90 (guidance on criteria); and para. 91 (the presumption in favour of preservation).

Having considered and dismissed arguments based on the alleged deficiencies of the Secretary of State's decision letter, Lord Bridges turned to the central issue:

> Perhaps the central issue in the (planning) appeal is the issue relating to planning policy. It was contended at the inquiry that the merits of a building proposed to be erected on the site of a listed building could not, as a matter of law, be a relevant consideration in considering whether to grant consent for the demolition of the listed building.

SAVE had abandoned this argument as a matter of law but, Lord Bridges continued, they had nevertheless adhered to the position that as a matter not of law but of planning policy,

> consent for the demolition of a listed building should never be granted if the building is still capable of economic use, that this policy is enshrined in paragraph 89 of Circular 8/87, and that the present applications for listed building consent, conservation area consent and planning permission should be regarded as a test case in relation to the question whether there can ever be any exception made to this policy.

After reviewing the arguments, the wording of the circular ('not very happily drafted'), and the Secretary of State's decision letter ('lacks the clarity and precision one would like to see'). Lord Bridges concluded that 'the policy states in unqualified terms in para. 89 is not intended to be absolute.' In particular, 'the fact that the circular spelt out the criteria against which applications for consent to demolition were to be considered, must imply that there could be exceptions.' Moreover the Secretary of State had explicitly said that his decision was taken 'in the special circumstances of this case.'

Finally Lord Bridges observed that

> The public controversy over this case arises from differences of opinion about traditional and contemporary architectural styles. These arouse strong feelings but are no concern whatever of the courts. It is a trite observation but can bear repetition in a case like this, that our concern is solely with the legality of the decision-making process, not at all with the merits of the decision.

The Moral

The moral that the DOE should have drawn from this excessively protracted case was not only that decision letters must be very carefully drafted (this was well known, although no draft is immune from legal casuistry), but that those preparing policy guidance need to avoid inserting passages that seem to offer unqualified doctrines or dogmatic statements that may appear to admit of no exceptions and which may unduly fetter the Secretary of State's statutory discretion in exercising his quasi-judicial appellate functions under the Planning Acts. Circular 8/87 certainly erred in that direction, but instead of taking the opportunity to put that right in PPG 15, the Department seems to have gone out of its way to reinforce the doctrine and to avoid scope for any discretion or flexibility in deciding cases on their merits rather than in accordance with doctrine. While that may please the arch-conservationists, it will not please those who are concerned for the long-term sustainability of conservation and planning policies. I return to that subject in the next chapter.

NOTES

The Departmental circulars referred to in this chapter were all published by HMSO.

1. Department of the Environment and Department of National Heritage (1994) *Planning Policy Guidance: Planning and the Historic Environment*, PPG 15. London: HMSO.
2. Secretary of State's decision letter 22 May 1985.
3. Inspector's report 12 October 1988.
4. Secretary of State's decision letter of 7 June 1989.
5. Save Britain's Heritage v. The Secretary of State for the Environment and No 1 Poultry Limited and city Acre Property Investment Trust Limited. Queen's Bench Division 19 December 1989.
6. Court of Appeal (Civil Division) 30 March 1990.
7. House of Lords, Judgement 25 February 1991.

23

Sustainable Conservation

Sustainability

We have heard much in the past few years about the need for 'sustainable development'. The common definition of this concept is 'to meet the needs of the present without compromising the ability of future generations to meet their own needs.'[1] It is a concept that is easier to endorse than to interpret and implement. But it informs much of current planning thought and activity. 'Conservation' in the sense in which it is used in this study has obvious relevance to this objective since by definition it conserves existing built assets and can facilitate their adaptation to new uses. Back in 1975 SAVE made the point: 'The fight to save particular buildings or groups of buildings is not the fancy of some impractical antiquarian. It is part of a battle for the sane use of all our resources.'[2] The current concern for sustainability gives a new validity to conservation policy.

As with other aspects of sustainability, however, it is a concept that, if carried to extremes, may not result in an effective use of resources. To insist on conservation as the primary objective of policy, in pursuit of sustainability, may impose unreasonable costs on the present, may defer problems to the future, and may impede renewal and creativity that are necessary to a sustainable future.

While conservation can serve the interests of sustainability, the question now is whether the system of conservation that we now have is itself sustainable.

I suggest that we should think in terms of 'sustainable conservation'. That would involve an approach to conservation that preserves the best of the heritage but does so without imposing insupportable costs and which effects a rational balance between conservation and change. In chapter 13 I quoted from Roy Worskett's book of 1969: 'We must understand physical change and its basic social and economic origins. We must be clear what we want to conserve, and why, and as part of the process of town planning we must develop techniques to cope with such a fundamental conflict.' Today one would modify that statement by recognizing that, in so far as such a 'conflict' exists, it may not be 'fundamental', since the concept of sustainable conservation can help to contain it.

This study of the political and legislative history of conservation has shown that the statutory system of conservation has been sustained and developed over the past fifty years in a quite remarkable way. Despite successive deregulatory initiatives, both the town planning system and the instruments of conservation have been strengthened rather than weakened. The abolition of the GLC (which, with its predecessor the LCC had a good reputation in conservation) and the metropolitan counties was damaging, but there has since been a modest revival of regional planning incorporated in the DOE's regional planning guidance. The current changes in local government organization (far less extensive and disruptive than they might have been) may stimulate new efforts at coordination at the subregional level as the counties, having largely survived reorganization, are required to cooperate with the new unitary authorities. At the local level the unitary authorities have the opportunity to develop planning policies in their new development plans that incorporate more coherent conservation objectives.

While the mechanisms of conservation are available, and while many local authorities have a good record in practical conservation work, there is far less sense of purpose at the national level and much uncertainty about the direction that conservation should take in the longer term. PPG 15, which the government intended to provide the definitive statement of conservation policy, is in fact for the most part a compilation of successive policy advice promulgated over many years.[3]

There are wider issues that have not yet been addressed. To his credit, Sir Jocelyn Stevens, Chairman of English Heritage, has been prepared to confront them, although he also observes that it is not for English Heritage to resolve them. In his speech to the Association of District Councils in May 1993 he concluded as follows:[4]

> At the end of the day it is neither possible nor desirable for English Heritage to set the agenda and priorities for the conservation of England's heritage. Fundamental questions such as :What do we mean by our heritage? What is priority and what is not? How much are we prepared to spend on conservation? raise complex and often conflicting cultural, economic and political issues which can only be resolved by debate at the national political level.

More than three years later it can hardly be said that these issues have begun to be addressed either by the politicians or by the public.

Attitudes to Conservation

We have now over 500,000 listed buildings, plus over 9,000 conservation areas which probably contain at least as many unlisted buildings covered by a similar protective net. A million buildings represents only about 5 per cent of the total building stock. But in many towns the town centre and much of the older neighbourhoods are covered by conservation areas. In Westminster, for example, nearly 90 per cent of the borough is in conservation areas and it contains some 12,000 listed buildings. Yet this is the centre of government, the cultural, commercial and entertainment centre of the capital city, and includes a substantial resident population. It cannot be immune from change, yet almost any change to any building (let alone demolition) requires listed building or conservation area consent. Many other historic towns and cities have imposed similar blanket conservation policies on themselves. Is this sustainable?

This study has shown that the British system of conservation has evolved mainly in response to growing public concern for the heritage. Sustainable conservation requires the support of public opinion. There is some evidence that attitudes may be changing or, at least, that doubts are being expressed about the extent and intensity of conservation policies.

If future historians seek a turning of the tide in attitudes to conservation in the last decade of the twentieth century, they may find it in the public response to the great fire at Windsor Castle in November 1992 and its aftermath.

Windsor Castle Fire

The fire at Windsor, which is one of the Queen's official residences and not an unoccupied Royal Palace, caused extensive damage to St George's Hall, the State Dining

Room, the Grand Chapel and two small adjacent rooms. The Queen's private apartments were not affected. In the immediate aftermath of the fire, the then Secretary of State for National Heritage, the Rt. Hon. Peter Brooke, rushed to the scene and promptly announced that all the damaged rooms would be reinstated, that the cost was likely to be in the order of £60 million and that the taxpayer would meet the bill. That precipitate statement prompted a vigorous controversy, which raised questions not only about what should be done at Windsor Castle and who should pay for it, but also prompted wider questions about conservation policy in Britain.

Although Windsor Castle was originally a Norman foundation, it had been changed and enlarged over the centuries. In the 1670s Charles II transformed much of it into a Baroque palace, to mark the restoration of the monarchy after the Civil War. Then in the 1830s it was rebuilt for George IV by Sir Jeffrey Wyatville in a variety of styles – Classical, Rococo, Gothic. The principal room, St George's Hall, was designed on the Gothic model or what was then termed the 'National Style'.

There were many who dismissed Wyatville's work as Victorian pastiche and argued that the opportunity should be taken not to restore the building to its previous state but to create something new and expressive of the late twentieth century, following the precedent of the 1670s and 1830s. In fact, as Giles Worsley, the architectural editor of *Country Life* was able to show, the quality of Wyatville's work was better and more interesting than generally assumed, and the extent of the damage not quite as devastating as originally thought.[5] Worsley wrote 'The damaged rooms are not bad Victorian pastiche but the grandest surviving interiors of one of our most cultivated monarchs.' Restoration was certainly a possible option. *Country Life*, however, acknowledged the possibility of alternative solutions and mounted a competition to explore the opportunities that the fire had created.[6]

At the same time, the distinguished architectural historian Mark Girouard, who had just completed a new history of the Castle, invited several architects to contribute ideas for rebuilding St George's Hall. He expressed his own view in an article in *The Independent*: 'If what decoration survives in the hall was the work of superb craftsmanship or had centuries of historic association, there would be an argument for keeping it and asking designers to work around it. But to treat Wyatville's surviving fragments as sacrosanct shows an absurd over-estimation of their quality and hamstrings any new design.'[7] Sir Hugh Casson, who had been President both of the RIBA and of the Royal Academy, and who had done some work at Windsor, described St George's Hall as 'the most boring room in Europe.'[8] So the experts were divided. The general public showed no great wish to see the building reinstated but the widely held view was that the Queen should pay for the rebuilding.

The results of the *Country Life* competition and of Mark Girouard's initiative were exhibited at the Architectural Foundation's gallery in London in July 1993. Some of the proposals were distinctly fantastical but some leading architects produced (at very short notice) striking ideas for combining the old and the new. In particular Richard MacCormac (former President of the RIBA) produced a stunning design for a new roof for St George's Hall which combined the principles of ribbed Gothic vaulting with high-tech materials and lighting.

Meanwhile, however, the Royal Household had assumed control of the rebuilding, with the Duke of Edinburgh in the lead. It was quickly decided to install a utilitarian new roof over St George's Hall, supposedly to protect the ruined interior. Then at the end of April 1993 the Lord Chamberlain announced (before the Architectural Foundation's exhibition had opened) that St George's Hall was to be 'restored as it was before', and that the Grand

Reception Room and the Green and Crimson Drawing Rooms were also to be restored. Only the Private Chapel and the two small adjacent rooms were to be redesigned. A limited design competition for the rebuilding was quickly mounted and the winning design that emerged within a year of the fire adopted a somewhat attenuated Gothic idiom, contrasting markedly with Wyatville's robust conception but conceding nothing to contemporary design.

Work is now in progress; it is expected to cost about £40 million and the castle will re-open to the public (no doubt with an extravagant entry fee) in 1998.

The outcome of the great fire at Windsor, and of the lively controversy and high hopes that it generated, was widely regarded as a great disappointment and a missed opportunity to achieve a new approach to conservation through creative design. Mark Girouard remarked 'we face a big yawn, meticulously evoked, at the cost of several million pounds.'[9] He had in fact anticipated such an outcome a few months earlier when he wrote 'I would find this a sad admission of failure, of a castle and monarchy withdrawn from the real world and perceived under glass as a tourist attraction.'[10]

Windsor Castle, St George's Hall by Wyatville (1829): before and after the fire of 1993.

The Heritage Industry

The perception of conservation as a vehicle for the tourism industry has caused much concern among many commentators. The most outspoken of these has been Robert Hewison, whose brilliant polemic *The Heritage Industry* attacks both what he regards as the unhealthy preoccupation with the heritage in all its forms and its commercial exploitation.[11] He portrays 'a country obsessed with its past, and unable to face its future . . . If the only new thing we have to offer is an improved version of the past, then today can only be inferior to yesterday. Hypnotised by images of the past, we risk losing all capacity for creative change.'

Hewison perhaps exaggerates the debilitating effect of what he calls the 'museum culture', but there is clearly a danger that the more meretricious forms of heritage enterprise may damage the historic fabric and mar the historical context. But the controls are available to prevent permanent damage to the fabric. Popular enthusiasm for the heritage generates funds for conservation and for the local economy, and also lends support for investment in preservation, which might otherwise seem a somewhat esoteric or elitist pursuit.

Maintenance and Repair

A more serious problem may be the growing backlog of repairs needed to the stock of historic buildings. English Heritage has reported that £150 million is needed for urgent repairs to the relatively few properties that are in its ownership. Its 1992 report on *Buildings at Risk* showed that, based on a sample survey, 37,000 of the total stock of listed buildings were in a vulnerable condition.[12] The Church of England faces an impossible burden of maintenance and repair, with a growing number of redundant churches, many of which are listed and among the most important features of the urban and rural landscape. Even the large sums generated by the National Lottery and available for the heritage will not go far to meet all these needs, especially if public opinion insists that part of that bounty should be used for less elitist causes, such as health care and education.

Given the scale of the problem, questions also arise about the kind of conservation measures that can be undertaken. The potential cost of thorough restoration can be formidable. The National Trust spent around £20 million repairing the damage to Uppark in West Sussex, which was burnt to a shell in 1989 (fortunately the cost was met by the contractor's insurers). The quality of the meticulous restoration work is quite astonishing but it will stand as a unique demonstration of craftsmanship, not as a model for general use.[13] It will be necessary to devise less expensive solutions to the problems of decay, and to accept that less exacting standards will need to be set.

At the other end of the scale, it must be recognized that it is not practical or reasonable to insist on applying the standards required in the best listed buildings to the vast number of houses and other buildings that are unlisted but in conservation areas. There has been some evidence of resentment at the over-zealous and pedantic requirements set by local conservation officers in policing repairs and improvements to such buildings. It can be argued that, if conservation area designation is to be effective, then the controls must be applied meticulously and with vigour. But too draconian an approach can impose unnecessary costs, inhibit necessary repairs, prevent useful adaptation, and consequently alienate public support for conservation.

Policy Review

The triumph of the conservationist cause over the past forty years must be tempered by awareness of the inherent problems that it presents. Conservation policy and its objectives have not been subject to fundamental review since the present system was established after World War 2, despite the attentions of successive Parliamentary Select Committees (see chapters 20 and 21). The Secretary of State for National Heritage has said that the government intends shortly to publish the results of its own review of the system.

It is not the purpose of this historical study to offer detailed prescriptions for the future of conservation policy. But there are some more general issues that could usefully be considered.

Uppark, West Sussex, by William Talman (1685–95): before the fire of 1989: restored by the National Trust.

These relate to questions of scope, selectivity, organization and responsibility.

Responsibility

Conservation is concerned with the management of resources, and, as with any aspect of management, it is essential to get the level of responsibility right. In practice, conservation operates at two levels – national and local; but at present these two levels are confused and not well defined. As a rule too much is expected of central government and too little responsibility is allowed to local government. The British system of planning and conservation is extraordinarily centralized and hierarchical. No other country in Western Europe or North America follows this pattern. Different arrangements are possible.

At present in England responsibility for conservation at the national level is shared between the Department of the Environment, the Department of National Heritage and English Heritage. A strong case can be made for reintegrating the conservation functions of the DNH with the DOE's responsibilities for urban and rural planning, environmental policy and local government. No benefit is gained from the present division of functions. But reintegration of functions does not mean that those functions should be as extensive as they are at present.

English Heritage should be retained as a Departmental agency (related to the DOE not the DNH) but its scope and functions need to be reassessed. In its first ten years or so it sought to establish itself as the prime mover in conservation and aspired to assert its role across the whole range of the built heritage. Inevitably it could not satisfy its own aspirations or the expectations it aroused. More recently, under new management, it has shown a far clearer sense of priorities and what its role should be. 'English Heritage cannot do everything' expresses the new realism, but the implications of this have not yet been fully resolved. English Heritage will do well if it ensures the good maintenance of those properties that are its direct responsibility, sustains its important archaeological work, and is able exceptionally to intervene to secure the protection of outstanding buildings that are in danger. It will probably need to take on a much larger share of the heritage of ecclesiastical architecture.

Listing

The process of 'listing', which over the past fifty years has become the foundation of the conservation system, is a reflection of its centralized nature and is also part of its strength. But the statutory definitions that underpin it, and the way in which the lists have been compiled, imply a uniformity of standards and selection which is misleading. It has also reinforced the common misconception that 'if it's listed it must be preserved' (and, all too often, 'if it's not listed, it ought to be'). The official line from the start has been that listing does not imply preservation but is simply a means of ensuring that the special architectural or historic interest of the building is considered and properly assessed when planning permission for alteration or redevelopment is sought. But, as we have seen, this objective approach was overtaken by the 'presumption in favour of preservation'.

The standardization and uniformity of the system derives from the statutory phrase 'buildings of *special* architectural or historic *interest*', and from the very rudimentary grading method adopted from the outset. With only about 6 per cent of the total inventory of listed buildings included in Grades I and II*, the remainder now all fall into the amorphous Grade II. It seems improbable that all those buildings could reasonably be described as being of 'special interest'. All buildings, even the most utilitarian, are of some interest to those with a sense of history and an awareness of the visual environment. To add the word 'special' implies discrimination but the implication remains that all listed buildings have been assessed in terms of uniform scholarly and aesthetic criteria and therefore warrant preservation, irrespective of other factors such as their significance in relation to the local environment or the practical implications of preservation. It is true that those factors can be taken into account when planning permission for alteration or demolition is sought, but that decision is likely to be based on the mere fact of its being listed rather than on a more complete assessment. Sometimes the fact that a building is not listed may result in the loss of a building or group of buildings that are a valuable component of the local scene and could still have a useful economic life. Sustainability should not be determined solely on aesthetic or scholarly criteria.

The listing system is now so well entrenched that it is probably impossible to start afresh. But decisions affecting such buildings, and responsibility for their conservation, might be more soundly based and effective if they were considered not primarily in terms of their 'special architectural or historic interest' but of their *importance* in cultural, aesthetic, historic, environmental and economic terms.

Thus buildings already listed Grade I and II* (the distinction does not appear to be significant) would be recognized as those of *national* importance, while those listed Grade II would be those of *local* importance. A substantial number of buildings at present in Grade II

De La Warr Pavilion, Bexhill, East Sussex, by Erich Mendelssohn and Serge Chermayeff (1933–36): one of the first inter-war buildings to be listed.

would justify 'promotion' to the higher category, since Grades I and II* reflect unduly exacting aesthetic standards and need not be so exclusive. The conservation of buildings in the first category should be regarded as a national responsibility and would be the primary concern of English Heritage. The costs of their conservation, in so far as that required public subsidy, would be met by central government. Those in the second category would be a local responsibility. It would be for local authorities, consulting local interests, to decide how far the costs of conservation should be met by local taxpayers. Public funds could be supplemented by private contributions and public subscription. An approach on these lines could, over time, result in a more rational distribution of responsibilities and a more effective use of resources.

The Secretary of State has announced his intention to publish a Green Paper canvassing possible changes in the listing system. That may again raise the question of prior notification, which was discussed when the 1944 and 1947 Acts were passing through Parliament. Some of the arguments advanced then, both for and against prior notification, are still relevant today. Spot listing procedures are now much more expeditious, and, if the building proposed for listing were in a conservation area, any attempt to demolish it would require consent. But there are new factors to consider. Since the great bulk of 'listable' buildings have now been listed, possible additions to the lists, especially of more modern buildings, may prove to be more contentious. Secondly, while in the past listing has generally been welcomed by owners as endowing their property (especially residential) with a certain cachet, that is now more questionable in view of the increased costs of maintenance and adaptation that this generally implies. Similarly, listing can effectively put a blight on commercial property and prevent redevelopment or renewal. Should these considerations be taken into account in deciding whether to list or are they only relevant when listed building consent is sought for demolition or alteration? Perhaps the time is ripe for a general review of the listing system and its related provisions. But it has worked well to date and any proposed changes to it should be intensely scrutinized.

Conservation Areas

Conservation areas present separate but related responsibilities. As we have seen in chapter 13, the 1967 act was drafted in terms of 'areas of special architectural or historic interest', and it was originally envisaged that those powers would be used to protect the setting of listed buildings. In practice, as the number and size of conservation areas have multiplied far beyond what was originally envisaged, such designations now often have little to do with special architectural or historic interest but everything to do with neighbourhood amenity and protection of the local environment. These are perfectly valid concerns but it suggests that it is not sensible to apply to these areas exactly the same types of control that apply to listed buildings. The fact that unlisted buildings in conservation areas are subject to virtually the same regime as listed buildings tends to overload the system.

While some conservation areas do contain large numbers of listed buildings and warrant the additional protection which that status confers, a great many contain very few listed buildings and sometimes none: respectable Victorian and Edwardian neighbourhoods, undistinguished but attractive shopping streets, the better class of inter-war suburban estates. The desire of local residents to protect their property and to achieve improvements in their community is wholly understandable. But those sentiments have little to do with 'special architectural or historic interest'. It would make more sense to treat such areas as a separate category and to develop methods more appropriate to their future management – traffic calming, parking control, landscaping, 'good housekeeping' measures. This approach would require full local participation and resident involvement in the work, but not the full panoply of conservation area controls. Nor should such action be confined to middle-class neighbourhoods: less well endowed areas may need it more and derive more benefit from it. Sustainable conservation should extend throughout the community.

Devolution

Under the present system, any refusal of planning permission or listed building or conservation area consent can be referred on appeal to the Secretary of State for the Environment. This results in matters of purely local importance being subject to review by an itinerant Inspector, appointed by the Secretary of State, who may have little knowledge of the local environment or of the value attached to it by local opinion. Such decisions should be taken at the local level, subject perhaps to intervention by the Secretary of State under reserve powers if an exceptional case arose where, for example, a listed building's importance had previously been undervalued. The local authority's decision could still be challenged in the Courts on grounds of illegality or natural justice; and cases of alleged maladministration could be referred, as at present, to the local Ombudsman.

There may well be doubts as to whether local authorities have the competence, expertise or resources to assume a larger role in conservation. Those doubts may be justified: the performance of local authorities in this field has certainly been uneven. But competence can be enhanced, and responsibility need not be devolved to the local level until competence can be demonstrated. The centralized State in Britain has become too dominant and a revival of local democracy and local responsibility is overdue.

The suggestions made in this chapter are not intended to be prescriptive, and they are certainly not intended to weaken the capacity for effective conservation. They are intended to strengthen local democracy, to encourage local interest in the local environment and to enlarge local participation in its protection and enhancement.

The Acts of 1990–91

In 1990 the accumulated planning and conservation legislation was consolidated in two new Acts – the Town and Country Planning Act 1990 (Eliz II 1990 c.8) and the Planning (Listed Buildings and Conservation Areas) Act 1990 (Eliz II 1990 c.9). While this arrangement made for more manageable legislative tomes, the separation of the two parts suggested a distinction between the planning and conservation systems whereas it is ever more important to ensure their integration.

Both these Acts were amended by the Planning and Compensation Act 1991 (Eliz II 1991 c.34), which in Section 26 amended Section 54/1990 so as to require that planning applications be determined in accordance with the development plan 'unless material considerations indicate otherwise'. This is the origin of what has become known as 'the plan-led system'. And so the endless legislative process continues.

The Future

While the statutory system may be in danger of overload and in need of review, the practice of conservation has markedly improved over the past twenty years or so. That improvement can perhaps be dated from the stimulus provided by European Architectural Heritage Year in 1975 (see chapter 15). Many local authorities have done excellent work, often associated with urban regeneration. Local organizations, encouraged by the Civic Trust, have developed in skill and resourcefulness. Long established bodies including the SPAB and the Georgian Group, together now with the Victorian Society and the Twentieth Century Society, provide an effective group of lobbyists. SAVE is still active. The National Trust's membership has grown dramatically, and English Heritage is building up similar support. Some private developers have taken on ambitious renewal schemes that incorporate major investment in the rehabilitation of older buildings. Individual architects, including some of the most distinguished, have brought outstanding creative skills to bear on the conservation and adaptation of listed buildings. These activities have shown that conservation objectives need not be incompatible with creativity and good design. These are all real assets that strengthen the cause of conservation.

This study of the history of conservation in England has shown how the policies and legislative structure have evolved over the past hundred years. There is scope for continued change of a kind that will help to ensure protection of the heritage and a creative future. This is the way forward for sustainable conservation.

NOTES

1. World Commission on Environment (1987) *Our Common Future*. Oxford: Oxford University Press, p. 8.
2. Binney, M. (1995) in *Perspectives*, May, p. 24.
3. Department of the Environment and Department of National Heritage (1994) *Planning Policy Guidance: Planning and the Historic Environment*, PPG15. London: HMSO.
4. Stevens, J. (1993) Speech to the Association of District Councils' Heritage Conference, 18 May.
5. Worsley, G. (1993) Windsor Castle, Berkshire. *Country Life*, 21 January, pp. 30–33.
6. Windsor Castle Competition. *Country Life*, 11 March 1993, p. 51 and 22 July 1993, p. 31.
7. Girouard, M. (1993) A Hall of Majesty, not fakery. *The Independent*, 6 May, p. 24.
8. Casson, H, quoted by Mark Girouard, (1993) *The Independent Magazine*, 10 July, p. 24.
9. Girouard, M. (1993) Room for improvement. *The Independent Magazine*, 10 July, p. 24.
10. Girouard, M. (1993) A Hall of Majesty, not fakery. *The Independent*, 6 May, p. 24.
11. Hewison, R. (1987) *The Heritage Industry*. London: Methuen, p. 137.
12. English Heritage (1992) *Buildings at Risk: A Sample Survey*. London: English Heritage.
13. Glancey, J. (1993) The high price of keeping up with 'heritage mania'. *The Independent*, 8 April, p. 14.

24

POSTSCRIPT: A CONSULTATION DOCUMENT

In May 1996, after the last chapter of this book had been written, the Department of National Heritage published *A Consultation Document on the Built Heritage of England and Wales.*[1] This glossy 61-page document reviews a range of possible changes to the listed buildings and conservation area procedures (and some relating to ancient monuments). For the most part these concern the minutiae of procedures and points of clarification in the legislation. The present study does not deal in detail with those procedures and therefore it is not necessary to recount these new proposals in detail.

In general, however, the consultation document demonstrates two factors. The first is the deregulatory paradox that in order to deregulate it is necessary to regulate. However irksome a regulation may appear to the deregulators, it may not be practical to abolish it entirely. But, if it is to be modified or reduced in scope, it will be found necessary to enact new regulations redefining its ambit; and the last state may be worse than the first. Amended regulations may generate new complexities and uncertainties. Thus, in the new consultation document, many possible changes are canvassed and rejected, or offered with marked reservations.

Secondly, the fact that so few significant changes are proposed demonstrates that statutory conservation has driven itself into something of an impasse. Any change, even the most benign, is liable to be greeted with suspicion by committed conservationists, and any support that it attracts from other interests (such as developers or architects) reinforces that suspicion. We have seen the same phenomenon with Green Belt policy. Any proposal for a more intelligent policy, or one more responsive to local conditions, is greeted with such howls of anguish that the politicians rush into reverse and may end up promulgating an even more inflexible policy in order to demonstrate their environmental correctness.

Thus it is with the new consultation document. It begins with a somewhat laboured manifesto protesting the government's commitment to 'Protecting the Heritage' (the title of the document) and thereafter makes no attempt to review or redefine policy. There are, however, a few points of interest that might lead to changes or improvements in the present regime. We can briefly review these.

The first issue is whether proposals to add buildings to the statutory list should be subject to wider consultation than at present. As we have seen in chapter 4, the question of notifying owners of proposals to list their property was raised during the Parliamentary debates on both the 1944 and 1947 Acts, but the government decided against it. The present consultation document reaches the same conclusion, mainly for reasons of cost. The government also rejects wider public consultation and offers the stern observation that 'Comments that amount to no more than an unsupported expression of personal views on or feelings about the building do not in the Government's view serve a useful purpose and there is no intention simply to count votes.' But two exceptions are made: general consultation is to be offered in the case of buildings recommended for listing by English Heritage in the light of their 'thematic studies' (i.e. studies of certain building types that are known to be under-represented in the lists), and also in the case of proposals for spot-listing buildings less than fifty years old. It is felt that these two categories will include 'those modern buildings on whose merits there may be no immediate consensus and on which a wider debate may therefore be particularly helpful.' English Heritage has recently initiated a consultation process on the listing of modern (mostly post-1945) buildings.[2]

The new document also canvasses the possibility of listing buildings on a 'provisional' basis while consultation on prospective listing is in progress. This is seen to raise some difficulties (e.g. in relation to compensation). But the possibility is also raised of 'provisional' listing in a different sense – i.e. to list selected modern (post-1940) buildings in a way that affords them full statutory protection but which is explicitly subject to review after ten years – at which time the building's listing would either be confirmed or deleted. It is suggested that such buildings should be identified in the lists by the letter 'M' (presumably meaning 'modern' – or possibly 'maybe'). This seems a sensible way of extending the benefits of listing to modern buildings and in a way which may lead to more such buildings being listed than if they were immediately added to the 'permanent' list. But that proposal itself is likely to attract controversy.

The document also raises the interesting question of whether, in deciding whether or not to list a building, the Secretary of State should 'take into account reasons other than architectural or historic interest, in particular economic, financial or personal considerations.' However, the government sees major difficulties in this suggestion and 'will not easily be persuaded that a change of approach would be desirable.' No change is the easy option and there are good arguments for adhering to the long established policy that the case for listing should be considered solely on the basis of a building's special architectural or historic interest, and that the case for demolition or redevelopment should be considered only when specific proposals are made that require listed building consent and/or planning permission. The difficulty with this logical approach is that, as we have seen, the fact that a building has been listed is now widely assumed to mean that it should be immune to demolition in perpetuity. Thus the case for renewal or redevelopment is prejudiced in favour of no change from the outset.

Certain useful changes are proposed in relation to conservation areas. The first is that local authorities should be required (not merely encouraged as at present) to carry out a 'reasoned appraisal' of any potential conservation area prior to formal designation; and also that the designation must include a statement 'identifying the specific features of the area that it is considered desirable to preserve or enhance.' Such a statement, it is suggested, 'would act as a benchmark against which future proposals for development or demolition could be judged'; and it could also serve 'as an authority's first step towards carrying out its statutory duty to formulate and publish proposals for the preservation and enhancement of its conservation areas.' It must be said that the latter statutory requirement is at present honoured more in the breach than in its observance, but these proposals are to be welcomed as providing a more coherent and positive approach to conservation areas.

The document also canvasses possible modifications to the control regime affecting unlisted buildings in conservation areas. The first, which the government seems disposed to accept, is that conservation area consent should not be required for demolitions affecting the interior of a building but only for works that affect its external appearance. But the government is not persuaded that the requirement for conservation area consent should be 'relaxed still further' so that consent for demolition should be required only where that would affect the appearance of 'the area itself' as distinct from the individual building. In rejecting this notion the government notes that conservation areas are 'particularly susceptible to creeping development'. That is certainly a constant danger but it suggests that even the current barrage of controls tends not to offer very effective protection in the absence of detailed policies for an area and effective enforcement. Those are matters that call for the more purposeful use of existing powers rather than new ones.

After listing some detailed improvements to the Ancient Monuments legislation and modifications to Crown immunity, the consultation document ends by proposing that English Heritage should be allowed to discard its cumbersome official title of 'The Historic Buildings and Monuments Commission for England'. And finally, in accordance with the spirit of the age, that English Heritage should be empowered to charge for its services, with the exception of those cases where it is required to be consulted – and in the case of services rendered to government departments.

In sum, the consultation document represents a worthy attempt to overhaul the present statutory system of conservation, but it steers well clear of any wider policy matters. There is no attempt to address some of the issues raised earlier in chapter 23 and in preceding chapters. These include questions concerning the scope and purpose of conservation; the balance between the interests of preservation and the case for renewal and innovation; the need for such a strongly centralized system; the case for distinguishing between buildings and areas of national importance and those of local importance; the need to redefine the concept of conservation areas so as to recognize that in many areas the case for their protection resides in qualities other than 'special architectural or historic importance'; questions of funding and the demand that conservation can reasonably place on national and local resources; and the need to resolve the uneasy division between Ministerial responsibilities for planning and conservation. These are all matters which, as we have seen in the course of this study, have been raised in the past and will continue to be raised again.

History is an underused resource in public administration. My hope is that this history of conservation policy over the past hundred years or so will be of some assistance to those who care for the future of the built heritage and who also care for the interests of renewal, innovation and creativity.

NOTES

1. Department of National Heritage (1996) *Protecting Our Heritage A Consultation Document on the Built Heritage of England and Wales*. London: HMSO.

2. English Heritage (1996) *Something Worth Keeping? Post-War Architecture in Britain.* London: English Heritage. This provides an overview of the subject and was published to coincide with a 'programme of public consultation' on the listing of post-war buildings (including schools, council housing, churches and industrial buildings) which was to last from March to September 1996, including an exhibition at the RIBA where visitors were invited to record their view on the case for listing specific buildings – both those proposed by English Heritage and any others.

Appendix A: Chronology

This brief chronology lists notable dates in the policy history of urban conservation and the chapters in which the main reference to them occurs.

		chapter
1717	Society of Antiquaries founded	2
1839	Camden Society founded	2
1845	Ecclesiological Society founded	2
1849	Ruskin : *The Seven Lamps of Architecture*	2
1877	Society for the Protection of Ancient Buildings (SPAB) founded	2
1882	Ancient Monuments Protection Act	3
1894	Committee for the Survey of the Memorials of Greater London established	3
1895	National trust founded	3
1908	Royal Commission on the Ancient and Historical Monuments and Constructions of England (RCHM) established	3
1909	Housing, Town Planning, Etc. Act	4
1913	Ancient Monuments Consolidation and Amendment Act	4
1923	Housing Etc. Act	4
1932	Town and Country Planning Act	4
1937	National Trust Act	9
1944	Georgian Group founded	5
1944	Town and Country Planning Act	7
1944	Advisory Committee on Listing	8
1944	*Instructions to Investigators*	8
1947	Town and Country Planning Act	8
1948	Gowers Committee (report 1950)	9
1953	Historic Buildings and Ancient Monuments Act	9

		chapter
1953	Historic Buildings Council for England established	9
1957	Civic Trust founded	12
1958	Victorian Society founded	10
1958	Archbishop's Commission on Redundant Churches (report 1960)	16
1962	Local Authorities (Historic Buildings) Act	10
1967	Civic Amenities Act (Circular 53/67)	13
1967	*Historic Towns: Preservation and Change*	13
1968	Town and Country Planning Act (Circular 61/68)	13
1968	Redundant Churches Fund established	16
1968	*Four Studies in Conservation*	13
1970	Department of the Environment established	14
1971	Town and Country Planning Act (Consolidation)	14
1972	Town and Country Planning (Amendment) Act (Circular 86/72)	14
1973	Circular 46/73 *Conservation and Preservation*	14
1974	Circular 102/74 *Historic Buildings and Conservation*	14
1974	Town and Country Amenities Act (Circular 147/74)	14
1975	European Architectural Heritage Year (EAHY)	15

		chapter
1975	SAVE Britain's Architectural Heritage founded	15
1979	Thirties Society founded (now Twentieth Century Society)	23
1980	Faculty Jurisdiction Committee (report 1984)	16
1981	*Organisation of Ancient Monuments and Historic Buildings in England* (consultation paper)	18
1982	*The Way Forward*	18
1983	National Heritage Act	18
1983	English Heritage established	19
1986	Environment Committee Enquiry (report 1987)	20
1987	Circular 8/87 *Historic Buildings and Conservation Areas: Policy and Procedure*	22
1990	Care of Cathedrals Measure	16
1992	Department of National Heritage established	21
1992	National Audit Office report: *Protecting and Managing England's Heritage Property*	21
1992	*Managing England's Heritage: Setting Priorities for the 1990s*	21
1993	Public Accounts Committee report	21
1994	National Heritage Committee report: *Our Heritage: Preserving It: Prospering from It*	21
1994	Planning Policy Guidance 15: *Planning and the Historic Environment*	22
1996	*Protecting Our Heritage*. A Consultation Document on the Built Heritage of England and Wales	24

Post-war Governments

from	*to*	*Party in power (Prime Minister)*
Aug. 1945	Feb. 1950	Lab. (Attlee)
March 1950	Oct 1951	Lab. (Attlee)
Oct 1951	May 1955	Cons. (Churchill)
June 1955	Sept 1959	Cons. (Eden, Macmillan)
Oct 1959	Sept 1964	Cons. (Douglas Home)
Oct 1964	March 1966	Lab. (Wilson)
April 1966	May 1970	Lab. (Wilson)
June 1970	Feb. 1974	Cons. (Heath)
March 1974	Sept. 1974	Lab. (Wilson)
Oct 1974	April 1978	Lab. (Wilson, Callaghan)
May 1979	May 1983	Cons. (Thatcher)
June 1983	May 1987	Cons. (Thatcher)
June 1987	March 1992	Cons. (Thatcher, Major)
April 1992		Cons. (Major)

Appendix B: Instructions to Investigators

Extracts from the 'Instructions to Investigators' issued in confidence in 1944 by the Ministry of Town and Country Planning to those preparing the first statutory lists of buildings of special architectural or historic interest under section 42 of the Town and Country Planning Act 1944.

The Varieties of Special Interest

The Act speaks of special architectural and historic interest and any building to be listed may have both, but must have one or the other kind of interest. Of course in a great measure they coexist. Most of the buildings which interest the architect also interest the historian and conversely, but the two kinds of interest combine in very different proportions and ways, between the extreme cases where the one or the other only is in question. Under each head, the historical and the architectural, several distinct approaches or criteria can be recognised which it would certainly not be easy but is probably not necessary to reduce to common terms. So long as a building has special interest from any of the following points of view it can properly be listed or at least submitted for listing, since the lists put in by investigators will undergo a certain degree of censorship at Headquarters.

The first and clearest case is that of the building which is a work of art, the product of a distinct and outstanding creative mind. There can be no doubt that every such building should be listed and that effects should be made to preserve every one of them. This class includes not merely the greater national monuments but such minor masterpieces as Lindsey House, Lincoln's Inn Fields; the Jacobean Hunting Lodge at Sherborne, Gloucestershire; Abingdon Town Hall; the Church of St Mary Woolnoth; the Custom House, King's Lynn, 44 Berkeley Square; Wick House, Richmond, Surrey; or 1 Palace Green, Kensington.

The next type is that of a building which is not a distinct creation in this sense but possesses in a pronounced form the characteristic virtues of the school of design which produced it, such as Eagle House, Mitcham; Cheyne Row, Chelsea; or the Paragon, Blackheath.

A third type of building which falls into neither of the above classes yet may qualify for listing on aesthetic grounds is the outstanding composition of fragmentary beauties welded together by time and good fortune, such a building as St James's Palace; Bisham Abbey; or the Deanery at Winchester.

Whether the interest of a building whose

importance derives from its place in the history of architecture should be called architectural or historical or both is a matter of no importance but there probably are buildings which fall into this class and no other, buildings whose sole but real value is that they exemplify a link in the chain of architectural development which if they were to perish would not be represented or represented so well. Nicholas Barbon's houses in Bedford Row, Strawberry Hill, Dance's Guildhall front, or Thackeray's house in Palace Green, Kensington, might be examples.

In this class should be included those architectural freaks which have sufficient character to be of interest, such as the triangular lodge at Rushton, the pagoda at Kew or the Tattlingstone Wonder.

It must be understood that Architectural History for our purposes includes not only the history of architectural design but equally the history of structural, including engineering, technique, and that for our purposes a steel bridge is as much a building as a cathedral. Certain industrial buildings are landmarks (whether we call them architectural or historical makes no matter) of the mechanical and industrial revolution , and thus ought certainly to be listed, though it may be that the investigators will wish to seek specialist advice in the matter. Examples are the iron bridge at Ironbridge, the Colebrookedale Ironworks, Burton's Conservatory at Kew and the roofs of St Pancras Station and Lime Street Station, Liverpool.

At this point attention should be drawn to the absence from our terms of reference of any lower limit of date. We may list buildings down to the present moment but we must of course be increasingly selective as the present day is approached. It may be said very roughly that down to about 1725 buildings should be listed which survive in anything like original condition. Between that date and 1800 the greater number of buildings should probably be listed though selection will be necessary. Between 1800 and 1850 listing should be confined to buildings of definite quality and character. From 1850 down to 1914 only outstanding works should be included and since 1914 none unless the case seems very strong and it appears possible that the building may not be brought to light by central research. It is, however, desirable that the selection of buildings for the last 150 years should comprise without fail the principal works of the principal architects and to some extent it may be possible to secure this by central research. The results will, however, have in every case to be sent to the local investigator for checking on the spot.

Coming now to the pure historic interest, it is possible to distinguish two main aspects which may be called the evidential and the sentimental. The Gothic ruins of abbey or castle, a matter of sentiment only to the tourist, are valuable to the historian as first class evidence of the religious, social or military organisation of their period, and so with the dovecote, the windmill, the warehouse, the Martello Tower or the recent pill-box. It is obvious that as evidence for history practically all old buildings are of direct and substantial value, the slums as much as Mayfair, and further, that the substantial preservation of a whole village, say, is more helpful than that of the same number of houses spread over a county. On the other hand it is obviously essential to be selective.

Under this heading the sociological interest of buildings, only now beginning to be seriously studied, has a very important place. There are, in every part of the country small buildings, particularly farmhouses, of regional style, reflecting the yeoman's activity and life in the sixteenth and seventeenth centuries; a wide range of characteristic examples of these should be preserved. In the more remote and thinly populated regions such as the Welsh and Northern Uplands, moreover, examples survive unaltered of cottages and farmhouse type which preserve the living and working arrangements characteristic of very remote periods, often indeed periods much more remote than those of the actual erection of the building in question since a Welsh seventeenth century building may reflect arrangements prevalent in England far back in the Middle Ages. It is not possible that the present urgent listing should include the archaeological investigation necessary to map out the distribution and nature of these regional types on which alone a final assessment of the relative importance of their examples can be based. But the investigators ought at any rate to be aware of the existence of this problem and in so far as they can recognise a regional type, map its distribution and note its best examples and they will add conspicuously to the value of our material. It should be added that from this point of view unaltered original condition is a factor of special importance. It is, of course, well understood that this archaeological interest must often be at war with the interest of the adaptation of buildings for modern living. Indeed the neglected or decayed building which the Medical Officer of Health is most anxious to condemn will often be that which interests the archaeologist most. This conflict is not one which investigators will be called upon to attempt to solve. But if they will say what buildings are of interest and in what degree, they will have afforded part of the basis on which others must attempt to solve it.

Under the statute, as soon as possible after listing, the Minister has to serve a notice on every owner or occupier of a listed building. For this and other obvious reasons of administrative convenience the normal unit of listing will be the individual dwelling house or other property, but it may often happen that the unit of architectural or historic interest is not the individual house but the whole context of which it forms a part, e.g. a classical terrace of houses such as the Royal Crescent, Bath, is a single architectural unit and should be listed and graded as such. This point is of especial importance for many planned streets and terraces of the first half of the nineteenth century in which it might be difficult to say that any single house was of special architectural interest though it is quite certain that the street as a whole does and will, so long as it can be preserved as a whole, possess such interest. This planned architectural group is the first type of group that has to be borne in mind.

The next is what may be called the accidental or pictorial architectural group where, by the good fortune of architectural good manners or a prevalent unity of feeling and approach at the time and place of building, a row of separately planned and built houses blend together into a group which in its wholeness gives a greater value to many of its members than they would have if they stood alone. Such groups of seventeenth, eighteenth and early nineteenth century buildings will often be seen in the High Streets of ancient market towns. In such a context it will often be desirable for the sake of its value to the group to list a building which might not be listed if it stood by itself. There is a nice distinction to be observed here; a building must not be listed merely because its neighbours are good and one is afraid that if it were to be demolished it would be replaced by some vulgar monstrosity which would not be tolerable, and least of all in good company. The design of new buildings can and should be controlled under general planning powers and it is not therefore proper to place on the statutory list a building under section 42 simply to ensure a congruity of neighbourhood which could as well be achieved in a new building. it must be possible to say that the old building, however plain and ordinary in its kind, has some quality in relation to the context which no new building could have.

There will probably be difficult borderline cases where investigator's feeling is that what is worth preserving somehow is a general character, in a street, town or quarter of a town, which is the cumulative effect of the grouping or repetition of a type of building, of which it is hard to say that any single specimen is more important than any other and yet it is certain that some, if not too many, could without real loss be spared. It is the Ministry's view that, while such buildings as these ought to receive special consideration and protection, this should be given rather by simple notice from the Ministry to the Local Authority and by the normal exercise of planning control than by Statutory listing under Section 42 of the Act. Investigators are therefore asked to exclude from their draft Statutory Lists but to note on *Supplementary Lists* buildings which have in their view cumulative group or character value, but which have not that degree of intrinsic architectural or historic interest which would naturally be called special interest. Group interest is not excluded from the Statutory Lists but the test set out above must be applied to it.

The group whether for *statutory* or *supplementary list* purposes need not necessarily be a street group. A group of cottages round a green or in the open country comes as properly under this heading; and under it may be brought also the building whose value lies in the part it plays in a landscape. The more usual and obvious case of this kind is that of the folly, obelisk, ruin and the like deliberately designed to play a part in the landscape composition. Where the avenue or vista that building was designed to terminate still exists, the building retains a value in that relation, regardless of any other consideration whatever, and may for that reason alone deserve listing.

A rarer case is that of the building which, by happy chance, contributes to the natural landscape. It is not positively asserted that there are cases of this kind in which listing is proper, but they are certainly not ruled out *a priori*, and if an investigator feels that he can make a case for listing on this ground he should do so.

The sentimental interest is more elusive and yet sentiment is probably the strongest single thread in our interest in the past. This may be entirely specific, arising from a particular event or a particular person, e.g. the scene of the Rye House plot or Keat's house at Hampstead; or less specific as connected with a class of well-known persons or a succession of events, e.g. the Albert Hall, or the Stock Exchange; or merely general, e.g. the feelings excited by prehistoric remains, Roman roads, or mediaeval abbeys. This historic sentiment is at once a very complex and comprehensive feeling. At its lowest it is an aspect of national self-respect, weak in early societies which rapidly destroy their past, but growing ever stronger with the lengthening of time, and most particularly in a society like ours whose historical development has not been catastrophic. The older it grows the more it looks back to its youth. But, of course, it can be a great deal more than this, old buildings having the power to kindle the historical imagination in a way denied to documents which supplement them.

Where such interest consists in a personal association, the first question which must be asked is how well or really that association is authenticated. Local tradition is not always a trustworthy guide and where investigators, having examined the evidence available to them, still remain doubtful, they should put down what they know and refer the matter to Headquarters. There are cases, however, where the continuing belief in a tradition may itself have constituted an historic interest, even though the original tradition may be baseless. This is not the place to enter upon discussion of the authenticity of Shakespeare's birthplace at Stratford-on-Avon but it can unhesitatingly be said that even were it proved absolutely that Shakespeare had never entered the house, nevertheless the existence since Garrick's time of an intensive cult centering upon that house would constitute an historic interest amply sufficient to place the house in the first class of the lists.

A similar argument may perhaps in certain cases justify the listing of buildings whose association is with characters or events in fiction, such as the birthplace of John Halifax, Gentleman, at Tewkesbury or the White Horse Inn at Ipswich where Mr Pickwick's room is shown.

Again it should be asked how important in national or local history was the person in question and how close was his association with the building? In particular, how far is it likely that he was conscious of the building as a factor in his life? The Bronte's life in Haworth Parsonage is evidently of the first importance as a dominating element in their experience reflected throughout their work. Less dominating and intense, but not unlike in kind, are the associations of Blake with Felpham, Ruskin with Denmark Hill, Darwin with Downe, Dickens with Gadshill or Tennyson with Farringford. But it is difficult to feel that the same interest attached to the transient and little remembered associations of lodgers or tenants upon short lease, however distinguished with town houses in London or Bath. This kind of association is not to be ignored but they should be looked at more critically.

A special kind of interest perhaps attaches where the historical personage either built or substantially altered or extended the house in question. In such a case it may bear upon it physical marks of his mind and taste which are interesting. Thomas Hardy's house, Max Gate, is a case in point; Sir John Soane's house would be another were it not obvious that there is a very positive and overriding architectural interest. Associations with historical events and phases of historical development may afford good grounds for listing, such as those of the Ship Hotel at Greenwich where the whitebait dinners were held, the manager's house at Vauxhall gardens, now St. Peter's Vicarage, or the booking office of the Stockton and Darlington Railway at Stockton.

An instance may fall between this category and that of sociological interest is that of the houses on the Charterville allotments near Witney, the fragmentary relic of a projected Chartist Utopia.

Finally it is necessary to draw attention to the special value of whole groups from the historical point of view. The preservation of the character of a whole town such as Conway or of the eighteenth century character of Bath has a historical value almost of a different order from that attaching to the preservation of individual houses within these groups. From other points of view a single eighteenth century building of moderate quality acquires extra value from being placed in an otherwise uninteresting wilderness of bleak modernity, but from this particular historical point of view the case is reversed and the maxim is 'to him that hath shall be added'.

An attempt like this to express in words what cannot be so expressed must necessarily dwell disproportionately on abnormal and borderline cases for these, after all, are where difficulties occur; once the limits of our special interest are understood, what falls within them may be taken more or less for granted. But it should probably be said that the special cases discussed in detail in this chapter will provide only a fraction of the total number of listed buildings. The great bulk and staple of the work will deal with clear and undoubted examples of fine buildings and it may be guessed that numerically the eighteenth century will (in total, though not in all localities) have a clear preponderance over any other.

The Field Technique of Listing

It is not necessary and will not always be desirable that at the early stages of their work investigators should have any contact with the local authority for the areas which they are listing. At a later stage, it will, of course, be necessary to go to these authorities for the names of owners or occupiers of the listed buildings, and it is, of course, possible that during the actual process of listing particular local authorities could and would give information of great value to investigators which might materially shorten their work. This, however, will differ from one case to another. Before approaching any Local Authority for the first time investigators should consult the Regional Planning Officer. Before he starts in any area the investigator should arm himself with all the existing local information which may save him time. He will generally be aware of any local archaeological and other societies (if not Headquarters can help him) and will, through the Secretaries hear of individuals who may have knowledge which could help him. Local guide books and papers and perhaps old maps should be accessible through these channels and can obviously be of the greatest service.

For obvious reasons and especially since notices have to be served on owners of statutorily listed buildings it is of the greatest importance that each building should be accurately identified that this can only be done if the investigator marks each building on a map as he lists it. If he does this he will at least have the essential data from which at a later stage local authorities can by comparison with their rating lists and maps accurately ascertain ownerships. For the rural areas, ordnance survey maps on a scale of 6″ to the mile should normally be sufficient and will be supplied to investigators but for the towns and built-up areas 25″ maps will usually be necessary and likewise will be supplied. One inch Ordnance Survey maps can also be supplied. The lists have, under the statute, to be grouped by the areas of local planning authorities, which here mean urban and rural district councils and the councils of County Boroughs. Investigators should therefore begin by marking accurately on their maps the boundaries of the local authority's area which they are listing and should take care that they cover the whole of an area before sending in the list.

The lists served on local authorities under the statute will naturally be bare lists of buildings without description or comment, but for office use, it will be most important to have sufficient detail of each building to indicate in general terms what its character is and why it has been included in a list.

In order that the Ministry may have some guidance on the relative importance of different buildings and the degree of effort which ought in each case to be made to secure preservation in the face of any threat, every building should be graded I, II or III. The *Statutory Lists* will comprise the buildings in Grades I and II, the *Supplementary Lists* those in Grade III. In Grade I should be placed buildings of such importance that their destruction should in no case be allowed; in Grade II buildings whose preservation should be regarded as a matter of national interest so that though it may be that now and then the preservation of a Grade II building will have to give way before some other yet more important consideration of planning or the like, yet the Ministry will, in each case, take such steps as are in its power to avoid the necessity of this and where no conflict of national interest can be shown will take such positive steps as are open to it to secure the building's preservation. In Grade III will be placed (1) buildings of architectural or historical interest which do not, however, rise to the degree properly qualified as special, (2) buildings which so contribute to a general effect that the Planning authority ought in the preparation and administration of its plan to regard this effect as an asset worth trying to keep.

Where the *Supplementary List* is concerned with group effects rather than the individual character of buildings, the descriptions in it should in general be confined to such effects. It will, however, still be necessary to identify clearly, on maps and in descriptions, the buildings included.

The investigator may occasionally believe or know that a case may be made by someone else for the listing of a building which he would not himself include even in the *Supplementary List*. In such a case he should note it, but grade it IV, simply in order that the Ministry, if criticised for its exclusion, may have the material for an answer to the critic.

The investigator will find that in areas which are architecturally poor he will tend to lower his standards and, for instance, place in Grade II buildings which in another area he would consign to Grade III. This tendency is reasonable and should not be overcorrected. The strictly local value of a building can properly be given weight, and this may be enhanced by local rarity and relation to local character. For example, in the upland country of Wales and the North unsophisticated local types of construction acquire an importance which could hardly be conceded to comparable buildings in lowland counties of ancient prosperity, while in an otherwise bleak tract of modernity the survivors of an older order may acquire added importance merely because they strike an effective and Romantic contrast. In other words a grading related, as ours is, to the effort to be made for preservation rather than to intrinsic merit will shift with local circumstances its relation to intrinsic merit.

It is obviously important that standards for listing and grading should be more or less evenly applied by the investigators but it is not thought that this can be ensured by any other method than example, instructions on the site or the checking of actual lists. All these methods will, therefore, be used and members of the Headquarters staff will, for periods, accompany each investigator to explain the standards to him. Though, however, the application of these standards cannot be explained in words, their meaning can and it is important to grasp what has been said about them. The grades are not intended to express a simple assessment of architectural quality but on the contrary an assessment of the degree of effort which ought to be made to secure preservation in the light of a defined range of considerations. What are these considerations? They would not, of course, include general planning matters such as road or other alternative proposals for the use of the land on which listed buildings stand. Nor do they directly include such factors as the condition of dwelling houses in relation to public health. This is not a matter which it is expected that investigators will be competent to assess. But they do include the whole range of historic and architectural considerations, so that if, for example, neglect or bad occupation has so reduced the architectural interest of a building or group of buildings that they are no more than a relic of what they were, then the investigator should take this into account when deciding whether to list or not. Here, however, it is important not to be misled by a merely superficial appearance of neglect or poor decorative condition. All doubtful cases of this kind can be covered and made clear by explanation incorporated in the lists.

Appendix C: Listing Criteria

The criteria for listing buildings of special architectural or historic interest have been altered over the years since listing started in 1944. The criteria were first defined by the Maclagan Committee as described in chapter 8 and adopted by the Advisory Committee on Listing (the Holford Committee). The criteria were set out in more detail and somewhat amended in the *Instructions to Investigators* (see appendix B). Slightly different criteria were published in the report of the Ministry of Town and Country Planning 1943–51, as noted in chapter 10. It was not until 1974, however, that the criteria were published in a Departmental Circular – DOE 102/74. The relevant extract from that Circular is reproduced below, together with the equivalent extracts from DOE Circular 8/87 and from the current Planning Policy Guidance 15 of 1994.

It is noteworthy that while the Maclagan Committee and the *Instructions to Investigators* defined Grades I, II and III in terms of priority for preservation, Circular 102/74 simply explained the 'principles of selection' for listing without explaining the significance of grading or attributing to each grade its relative priority for preservation. Circular 8/87 repeated the principles of selection for pre-1939 buildings and added a further note on interwar buildings, and, for the first time in an official publication, set out the criteria for grading. Curiously, that Circular referred to priority for preservation only in relation to Grade II buildings, 'which warrant every effort being made to preserve them.' PPG15 sets out the current 'principles of selection' in an entirely new way, though incorporating some of the earlier language. It does not re-iterate the explanation of grading given in Circular 8/87; the only reference to grading is in a different part of the Guidance (paragraph 3.6) and is reproduced below.

One is left unsure whether these successive amendments to the principles of selection and the definition of grading were intended to have any distinct policy significance, or whether they represent merely some uncertainty about the system. The statutory provisions on listing make no reference to grading but require only the compilation of lists of buildings of special architectural or historic interest. Grading was introduced as an administrative device but the more recent revisions of the listing criteria suggest that the lists are essentially descriptive and that decisions on the preservation of individual buildings should be taken in the circumstances of the particular case.

Appendix to Department of the Environment Circular 102/74 of 8 July 1974

Listing of Buildings of Special Architectural or Historic Interest – Principles of Selection

I. On the first major revision of standards in 1967/68 the principles of selection were broadly as follows:

All buildings built before 1700 which survive in anything like their original condition are listed.

Most buildings of 1700 to 1840 are listed, though selection is necessary. Between 1840 and 1914 buildings must be of definite quality and character to qualify, except where they form part of a group, and the selection is designed to include among other buildings the principal works of the principal architects.

A start is now being made on listing a very few selected buildings of 1914 to 1939.

In choosing buildings, particular attention is paid to:

1. Special value within certain types, either for architectural or planning reasons or as illustrating social and economic history (for instance, industrial buildings, railway stations, schools, hospitals, theatres, town halls, markets, exchanges, almshouses, prisons, lock-ups, mills).
2. Technological innovation or virtuosity (for instance, cast iron, prefabrication, or the early use of concrete).
3. Association with well-known characters or events.
4. Group value, especially as examples of town planning (for instance squares, terraces or model villages).

II. In March 1970 it was decided to abolish the III grading and to up-grade to the statutory list certain categories of buildings that previously had been, or would have been graded III. A summary of these follows:

Types of Buildings to be Up-Graded to the Statutory List

(*a*) Small cottages of timber-framed construction dating from before about 1650–1700 previously thought to be too modest to merit statutory listing.
(*b*) Farm houses of the same period which have been re-faced in other materials but which show some trace of their ancient origin in individual features.
(*c*) In urban areas, timber-framed buildings so altered so as to appear largely c19 although clearly of ancient origin.
(*d*) Buildings of informal groups of mixed quality in which the groups were not previously thought so important that it was essential for each item to be statutorily listed.
(*e*) Buildings in a planned estate or group which have remained substantially intact but are of modest quality.

Types of Former Grade III Buildings to be considered on Their Merits and either Statutorily Listed or Classified as being of Local Interest

(*f*) Plain or very modest C18 or early C19 houses with few special features.
(*g*) Manor houses or mansions mainly of C18 date which have been so much altered in the C19 as to have more the character of that date than their original period.
(*h*) Buildings of any period whose ground floor has been converted into a shop and whose upper portion is not sufficiently impressive on its own to justify statutory listing.
(*i*) The last products of the Georgian tradition, viz: very late buildings of 1834–40.
(*j*) C19 buildings which have been badly mutilated by later alterations.

Appendix to Department of the Environment Circular 8/87 of 25 March 1987

Listing of Buildings of Special Architectural Interest or Historic Interest – Principles of Selection

The principles of selection for the lists were drawn up by the Historic Buildings Council (the functions of the former Historic Buildings

Council for England are now carried out by the Historic Buildings and Monuments Commission (HBMC)) and approved by the Secretary of State). They cover four groups:

1. All buildings built before 1700 which survive in anything like their original condition are listed.
2. Most buildings of 1700 to 1840 are listed, though selection is necessary.
3. Between 1840 and 1914 only buildings of definite quality and character are listed, and the selection is designed to include the principal works of the principal architects.
4. Between 1914 and 1939, selected buildings of high quality are listed (see below).

After 1939, a few outstanding buildings are listed.

In choosing buildings, particular attention is paid to:

1. Special value within certain types, either for architectural or planning reasons or as illustrating social and economic history (for instance, industrial buildings, railway stations, schools, hospitals, theatres, town halls, markets, exchanges, almshouses, prisons, lock-ups, mills).
2. Technological innovation or virtuosity (for instance, cast iron, prefabrication, or the early use of concrete).
3. Association with well-known characters or events.
4. Group value, especially as examples of town planning (for instance squares, terraces or model villages).

A Note on Interwar Buildings

The criteria for selecting buildings of the 1914–1939 period for listing cover two issues: the range of buildings which may be considered, and the quality of the individual buildings actually selected.

The criteria are designed to enable full recognition to be given to the varied architectural output of the period. Three main building styles (broadly interpreted) are represented: modern classical and others. The building types which may be considered cover nine categories, as follows:

(*a*) Churches, chapels and other places of public worship.
(*b*) Cinemas, theatres, hotels and other places of public entertainment.
(*c*) Commercial and industrial premises including shops and offices.
(*d*) Schools, colleges and educational buildings.
(*e*) Flats.
(*f*) Houses and housing estates.
(*g*) Municipal and other public buildings.
(*h*) Railway stations, airport terminals and other places associated with public transport.
(*i*) Miscellaneous.

The selection includes the work of the principal architects of the period.

Grading

The buildings are classified in grades to show their relative importance as follows:

GRADE I These are buildings of exceptional interest (only about 2 per cent of listed buildings so far are in this grade).
GRADE II* These are particularly important buildings of more than special interest (some 4 per cent of listed buildings).
GRADE II These are buildings of special interest, which warrant every effort being made to preserve them.
GRADE III A non-statutory and now obsolete grade. Grade III buildings are those which, whilst not qualifying for the statutory list, were considered nevertheless to be of some importance. Many of these buildings are now considered to be of special interest by current standards – particularly where they possess 'group value' – and are being added to the statutory lists as these are revised.

Extract from Department of the Environment and Department of National Heritage Planning Policy Guidance (PPG) 15 of September 1994

Principles of Selection

6.10. The following are the main criteria which the Secretary of State applies as appropriate in deciding which buildings to include in the statutory lists:

ARCHITECTURAL INTEREST: the lists are meant to include all buildings which are of importance to the nation for the interest of their architectural design, decoration and craftsmanship; also important examples of particular building types and techniques (e.g. buildings displaying technological innovation or virtuosity) and significant plan forms;

HISTORIC INTEREST: this includes buildings which illustrate important aspects of the nation's social, economic, cultural or military history.

CLOSE HISTORICAL ASSOCIATION: with nationally important people or events.

GROUP VALUE, especially where buildings comprise an important architectural or historic unity or a fine example of planning (e.g. squares, terraces or model villages).

Not all these criteria will be relevant to every case, but a particular building may qualify for listing under more than one of them.

6.11. Age and rarity are relevant considerations, particularly where buildings are proposed for listing on the strength of their historic interest. The older a building is, and the fewer the surviving examples of its kind, the more likely it is to have historic importance. Thus, all buildings built before 1700 which survive in anything like their original condition are listed; and most buildings of about 1700 to 1840 are listed, though some selection is necessary. After about 1840, because of the greatly increased number of buildings erected and the much larger numbers that have survived, greater selection is necessary to identify the best examples of particular building types, and only buildings of definite quality and character are listed. For the same reasons, only selected buildings from the period after 1914 are normally listed. Buildings which are less than 30 years old are normally listed only if they are of outstanding quality and under threat. Buildings which are less than ten years old are not listed.

6.12. The approach adopted for twentieth century listing is to identify key exemplars for each of a range of building types – industrial, educational, residential, etc. – and to treat these exemplars as broadly defining a standard against which to judge proposals for further additions to the list. This approach has already been successfully applied to the inter-war period, and English heritage is now engaged on a three-year research programme to extend it to the post-war period (subject to the '30 year rule' mentioned above). Proposals for listings in each building type will be made as each stage of the research is completed.

Selectivity

6.13. Where a building qualifies for listing primarily on the strength of its intrinsic architectural quality or its group value, the fact that there are other buildings of similar quality elsewhere is not likely to be a major consideration. But, as noted above, the listing of buildings primarily for historical reasons is to a greater extent a comparative exercise, and needs to be selective where a substantial number of buildings of a similar type and quality survive. In such cases the Secretary of State's aim will be to list the best examples of the type which are of special historic interest.

Aesthetic Merits

6.14. The external appearance of a building – both its intrinsic architectural merit and any group value – is a key consideration in judging listing proposals, but the special interest of a building will not always be reflected in obvious visual quality. Buildings which are important for reasons of technological innovation, or as illustrating particular aspects of social or economic history, may well have little external visual quality.

Historical Associations

6.15. Well-documented historical associations of national importance will increase the case for the inclusion of a building in the statutory list. They may justify a higher grading than would otherwise be appropriate, and may occasionally be the deciding factor. But in the Secretary of State's view there should normally be some quality or interest in the physical fabric of the building itself to justify the statutory protection afforded by listing. Either the building should be of some architectural merit in itself, or it should be well preserved in a form which directly illustrates and confirms its historical associations (e.g. because of the survival of internal features). Where otherwise unremarkable buildings have historical associations, the Secretary of State's view is that they are normally best commemorated by other means (e.g. by a plaque), and that listing will be appropriate only in exceptional cases.

National and Local Interest

6.16 The emphasis in these criteria is on national significance, though this cannot be defined precisely. For instance, the best examples of local vernacular building types will normally be listed. But many buildings which are valued for their contribution to the local scene, or for local historical associations, will not merit listing. Such buildings will often be protected by conservation area designation (see paragraphs 4.2 ff). It is also open to planning authorities to draw up lists of locally important buildings, and to formulate local plan policies for their protection, through normal development control procedures. But policies should make clear that such buildings do not enjoy the full protection of statutory listing.

Grading

3.6. The grading of a building in the statutory lists is clearly a material consideration for the exercise of listed building controls. Grades I and II* identify the outstanding architectural or historic interest of a small proportion (about 6%) of all listed buildings. These buildings are of particularly great importance to the nation's built heritage: their significance will generally be beyond dispute. But it should be emphasised that the statutory controls apply equally to all listed buildings, irrespective of grade; and since Grade II includes about 94% of all listed buildings, representing a major element in the historic quality of our towns, villages and countryside, failure to give careful scrutiny to proposals for their alteration or demolition could lead to widespread damage to the historic environment.

Appendix D: Statistics

There is only one regular source of statistics on conservation: *English Heritage Monitor*, which for the past seven years has been compiled and published by the English Tourist Board. In 1995 the *Monitor* was published jointly with English Heritage who provided financial support. The following selected statistics are taken from this source, with due acknowledgement. All the figures relate to England only.

1. Total Stock of Listed Buildings at 31 December 1994

Listed buildings	447,043
– of which Grade I	6,078
Scheduled Ancient Monuments	15,429
Conservation Areas	8,315
Listed Anglican Churches	12,970
– of which Grade I	2,959

2. Increases December 1993 to December 1994

Listed buildings	4,129
Conservation Areas	368
Scheduled Ancient Monuments	542

3. Distribution of Listed Buildings in 1991

Top Ten Local Authority Areas:

Cotswold	4,794
South Somerset	4,534
West Dorset	4,183
Westminster	3,689
North Wiltshire	3,604
Uttlesford	3,555
South Oxfordshire	3,438
Mid Suffolk	3,281
Chichester	3,176
North Cornwall	3,114

Top Ten Cities:

Bath	1,820
Huddersfield	1,751
Bristol	1,542
Liverpool	1,458
Birmingham	1,089
Oxford	1,029
Norwich	965
York	920
Exeter	875
Canterbury	778

4. Listed Anglican Churches

Top Ten Dioceses:

Lincoln	279
St Edmundsbury (Suffolk)	202
Norwich	198
Peterborough	168
Salisbury	134
Exeter	130
Ely	116
Chichester	109
Canterbury	106
Bath and Wells	105

5. Visitor Numbers 1994

Paid Admission:

Tower of London	2,407,115
Windsor Castle	1.900,000

Roman Baths and Pump Room Bath	871,308
Warwick Castle	755,700
Stonehenge	696,605
Shakespeare's birthplace, Stratford	591,205
Hampton Court Palace	543,061
Leeds Castle, Kent	537,965
Blenheim Palace, Woodstock	449,755
Buckingham Palace	420,000

Free Admissions

Canterbury Cathedral	2,250,000
Westminster Abbey	2,200,000
York Minster	2,000,000
Chester Cathedral	1,000,000
St Paul's Cathedral	700,000
Salisbury Cathedral	600,000
Winchester Cathedral	590,000
Norwich Cathedral	530,000
Buckfast Abbey, Devon	450,000
Exeter Cathedral	400,000

6. English Heritage Grants 1993–94 (£)

Secular buildings	11,816,540
Churches	12,458,934
Cathedrals	4,834,395
Conservation Areas	5,897,464
Town Schemes	4,579,798
Total	39,667,131

7. English Heritage Grants 1976–77 to 1993–94* (£)

Secular buildings	100,940,492
Churches	94,301,300
Cathedrals	10,862,155
Conservation Areas	63,052,462
Town Schemes	39,658,321
Total	308,814,730

*In 1984–85 responsibility for these grants was transferred from the Historic Buildings Council to English Heritage.

8. Listed Building Consent

Information on the depletion of the stock of listed buildings is sparse. The *English Heritage Monitor* for 1995 records that listed building consent was given for the total demolition of 92 listed buildings in 1993–94, including two Grade II*. But the rate of demolition appears to be declining. The 1993–94 rate was only half of that in 1988 and a third of that fifteen years ago when there were 301 demolitions. The number of consents given can be compared with the number of applications to demolish – 219 in 1994.

Accurate statistics in this field are impeded by problems of definition. Aside from total demolition, consent was given in 1994–95 for 'partial' demolition of 616 buildings; but this could include very minor works or the removal of incongruous later additions. Similarly, applications for consent to 'alterations' cover a wide range of work, since virtually any alteration to a listed building (or to an unlisted building in a conservation area) requires consent. There were about 9,000 such applications relating to listed buildings in 1994–95.

SELECT BIBLIOGRAPHY

This Select Bibliography lists books referred to in the main text. Official publications and periodical articles are referenced in the Notes to each chapter.

Ashworth, W. (1954) *The Genesis of British Town Planning*. London: Routledge and Kegan Paul.

Barker, F. and Jackson, P. (1990) *The History of London in Maps*. London: Barrie and Jenkins.

Binney, M. (1984) *Our Vanishing Heritage*. London: Arlington Books.

Brown, G.B. (1905) *The Care of Ancient Monuments*. Cambridge: Cambridge University Press.

Cherry, G.E. (1974) *The Evolution of British Town Planning*. London: Leonard Hill.

Clark, K. (1974) *The Gothic Revival: An Essay in the History of Taste*, 3rd ed. London: John Murray.

Croker, T.C. (1860) *A Walk from London to Fulham*. London: William Tagg.

Crossman, R.H.S. (1975) *The Diaries of a Cabinet Minister, Vol. 1: 1964–66* (ed. Janet Morgan). London: Hamish Hamilton and Jonathan Cape.

Cullingworth, J.B. (1993) *Town and Country Planning in Britain*, 11th ed. London: Unwin Hyman.

Dobby, A. (1978) *Conservation and Planning*. London: Hutchinson.

Ellis, C.W. (1928) *England and the Octopus*. Privately printed (new edition 1975, Blackie and Sons).

Esher, L. (1982) *The Continuing Heritage: The Story of the Civic Trust Awards*. London: Freney.

Evans, J. (1956) *A History of the Society of Antiquaries*. Oxford: Oxford University Press.

Eversley, D. (1973) *The Planner in Society: The Changing Role of the Profession*. London: Faber and Faber.

Fawcett, J. (ed.) (1976) *The Future of the Past: Attitudes to Conservation 1740–1974*. London: Thames and Hudson.

Ferry, B. and Sutcliffe, A. (eds.) (1983) *The Pursuit of Urban History*. London: Edward Arnold.

Harvey, J. (1972) *Conservation in Buildings*. London: John Baker.

Henderson, P. (1967) *William Morris: His Life, Work and Friends*. London: Thames and Hudson.

Hewison, R. (1987) *The Heritage Industry*. London: Methuen.

Hobhouse, H. (1971) *Lost London*. London: Macmillan (revised edition 1976).

Hobhouse, H. (1994) *London Survey'd: The Work of the Survey of London 1894–1944*. London: RCHM.

Hunt, L. (1857) *The Old Court Suburb*. London.

Hunter, M. (ed.) *Preserving the Past: The Rise of Heritage in Modern Britain*. Stroud: Alan Sutton Publishing.

Jenkins, J. and James, P. (1994) *From Acorn to Oak Trees: the Growth of the National Trust 1895–1994*. London: Macmillan.

Kain, R. (1981) *Planning for Conservation*. London: Mansell.

Keeble, L. (1948) *Principles and Practice of Town and Country Planning*. London: Estates Gazette.

Kennet, W. (1972) *Preservation*. London: Temple Smith.

Kosof, S. (1991) *The City Shaped: Urban Patterns and Meanings through History*. London: Thames and Hudson.

Kosof, S. (1992) *The City Assembled: The*

Elements of Urban Form through History. London: Thames and Hudson.
Lowenthal, D. and Binney, M. (eds.) (1981) *Our Past Before Us: Why Do We Save It?* London: Temple Smith.
Lowenthal, D. (1985) *The Past is a Foreign Country*. Cambridge: Cambridge University Press.
Murphy, G. (1987) *Founders of the National Trust*. London: Christopher Helm.
Nairn, J. (1955) *Outrage*. London: Architectural Press.
Pevsner, N. (1974) *Some Architectural Writers of the Nineteenth Century*. Oxford: Clarendon Press.
Roseburg, J.D. (1964) *The Genius of John Ruskin: Selections from His Writings*. London: Allen and Unwin.
Ross, M. (1991) *Planning and the Heritage: Policy and Procedures*. London: E. and F.N. Spon (second edition 1996).
Samuel, R. (1994) *Theatres of Memory*. London: Verso.
Smith, J.F. (1978) *A Critical Bibliography of Building Conservation*. London: Mansell.
Sproule, A. (1982) *Lost Houses of Britain*. Newton Abbot.
Suddards, R.W. and Hargreaves, J.M. (1996) *Listed Buildings: The Law and Practice of Historic Buildings, Ancient Monuments and Conservation Areas*, 3rd ed. London: Sweet and Maxwell.
Summerson, J. (1945) *Georgian London*. Harmondsworth: Penguin (second edition 1962, third edition 1978).
Summerson, J. (1948) *Heavenly Mansions and Other Essays on Architecture*. London: Cresset (new edition W.W. Norton, New York, 1963).
Sykes, C. (1985) *Private Palaces: Life in the Great London House*. London: Chatto and Windus.
Waterson, M. (1994) *The National Trust: The First Hundred Years*. London: BBC Publications.
Weideger, P. (1994) *Gilding the Acorn: Behind the Facade of the National Trust*. London: Simon and Schuster.
Worskett, R. (1969) *The Character of Towns: An Approach to Renewal*. London: Architectural Press.

INDEX

Adam, Robert 44
Adelphi, The, London, demolition of 44, 45, 84
Advisory Committee on Listing 62, 65–66, 67, 68, 69, 70, 73, 78, 80, 82, 86, 90, 133
aesthetic control 38
Albert Bridge, London 108
All Hallows, Lombard Street, London 49
ancient monuments 23–35, 37
Ancient Monuments and Archaeological Areas Act 1979 32
Ancient Monuments Board for England 136, 137
Ancient Monuments Consolidation and Amendment Act 1913 10, 30, 31, 119
  First Preservation Order 31
Ancient Monuments Protection Act 1882 1, 23, 25, 27, 29, 55, 119
Ancient Monuments Society 89
Archbishops' Commission of Redundant Churches 121–124
Ansell, W.H. 72
Archaeological Institute of Great Britain 12
archaeological societies
  development of 32
archaeology 2, 9, 12, 32
Architectural Association 21
Architectural Heritage Fund 113–114
Architectural Review 48, 49
Arts and Crafts Movement 32
Ashbee, Charles Robert 33
Ashworth, William 36
Association of District Councils' Heritage Conference 160
Aubrey, John 9
Averbury 23
Averbury, Lord, see Lubbock, Sir John
Avon, Earl of 138

Baldwin Brown, G. 25, 26, 28, 29
Balfour, Arthur 49
Banbury, Lord 40
Bank of England 44
Banks, Sir Joseph
  demolition of house of 49
Barry, Charles 16
Bath, *see also Four Conservation Studies* 114
Bedford Square, London 80, 81
Benson, Oxfordshire 81
Betjeman, Sir John 3, 82, 91, 112
Binney, Marcus 115, 125
Birk, Lady 133, 134, 138, 139, 140
'blue plaque' scheme 34
Blunt, Anthony F. 72
Bodley, G.F. 15
Boulting, Nickolaus 3, 9
Bridge, Lord 105
Bridges, Lord 121, 174, 175
  Commission of 127
British Archaeological Association 12
British Council on Archaeology 78
Britton, J. 12
Brooke, Peter 179
Buchanan, Sir Colin
  and Bath 98–99
building preservation orders 39, 40, 58, 59, 79–82, 101
Bunning, J.B. 83, 84
Burn, William 15
Burne-Jones, Sir Edward 20
Burrows, G.S.
  and Chichester 98–99
Butterwalk, Dartmouth 79
Buxton, Derbyshire 114
Byron, Robert 48, 49, 50

Camden, William 9
Camden Society 14, 21
Cantacuzino, Sherban 113
Cantrel, Timothy 133
Care of Cathedrals Measure 1990 128
Carew, Thomas 9
Carlton House Terrace, London 48, 49
Carlyle, Thomas 20
cathedrals 125, 128
  Durham 13, 14
  Exeter 63
  Lichfield 13
  restoration 13–15
  St Paul's 89
  Salisbury 13
Cathedrals Repairs and Grants Scheme 128
Central Advisory Board on Redundant Churches 124
Central Council for the Care of Churches 121
Channon, Paul 140
Chapman, Sydney 149, 150
Chatham, London 79
Chermayeff, Serge 185
Chenies, Buckinghamshire 112

Cherry, Gordon E. 133
Chester, *see also Four Conservation Studies* 113
Chesterfield House, London 46
Chettle, E. 65
Cheyne Walk, Chelsea, London 92
Church Assembly 121
Church of England 119, 120, 124, 125, 126, 182
churches 74, 119–129, 160
  alterations to 120–121
  and the Victorian restorers 13–15, 20
  listed Grade I 119
  London, destruction of 47
  post-Reformation desecration 9
  redundancy 121–124, 125, 126, 127, 129
Churches Conservation Trust 127
Chuter Ede, J. 73
Circular 53/67 100–101, 165, 167
Circular 61/68 101–102, 165–166
Circular 46/73 105, 166, 167
Circular 86/72 111, 166
Circular 102/74 106–107, 166, 167, 168, 201–202
Circular 147/74 107–108
Circular 8/87 165, 166, 168, 169, 170, 174, 175, 202–203
Civic Amenities Act 1967 64, 95–96, 97, 98, 100, 102, 165
Civic Trust 89, 111, 113, 150
Clapham, Sir Alfred 65
Clark, David 140
Clergy House, Alfriston, East Sussex 33, 34
Coal Exchange, London 82
  demolition of 83–84
Commercial Street, Leeds 114
Commission des Monuments Historiques 12
Commission for Ancient Monuments and Historic Buildings (England) 138, 139, 152
Commissioners of Works 23, 30, 31, 39, 40, 41
  duties of 31
Committee for the Survey of the Memorials of Greater London 34
Commonwealth, The
  House of Commons Order of 1641 9
Compton Verney, Warwickshire 80
compulsory purchase of buildings and monuments 29, 34
Conservation and Preservation, *see* circular 46/73
Consistory Courts 120, 121
Constitution of Otho 120
Consultation Document on the Built Heritage of England and Wales 189–191
Cornforth, John 11
Council for the Protection of Rural England 78
Council of Europe 111, 113
County Hall, London 145
Coutts Bank, London 108–109
Covent Garden 108–109
CPRE, *see* Council for the Protection of Rural England
Cranborne, Viscount 40, 55
Cranworth, Lord 40
Crosby, Theo 102
Crosland, Anthony 104
Crossman, Richard 89, 90–94
Crown Court, Godalming, Surrey 79
Crown Hotel, Yeovil, Somerset 79
Cullen, Gordon 86
Curzon, Lord 31

Dale, Anthony 69
Dartford, Countess of 111, 114
Davidson, Archbishop 121
Day, Alan 102
De La Warr Pavilion, Bexhill, East Sussex 185
Delafons, John 91
Department of National Heritage 2, 73, 138, 156–157, 162, 163, 183, 189
Department of the Environment 2, 73, 87, 95, 104, 124, 127, 133, 136, 137, 144, 156, 183
  Listing Branch 2
Despencer-Robertson, Major 40
Devey, George 15
Devonshire House, Piccadilly, London
  demolition of 50–51
Devonshire Place, London 80
Dibdin, Sir Lewis 121
Diocesan Advisory Committees 121
Directorate of Ancient Monuments and Historic Buildings 104
Dobby, Alan 3, 55, 64, 110
Dodsworth, Roger 9
DOE, *see* Department of the Environment
Dorchester House, London 46
Doughty Street, London 80
Dugdale, Sir William 9

ecclesiastical exemption 119–120, 124, 125, 126, 127, 128, 129
Ecclesiological Society 14
Eden, W.A. 66
Edinburgh, Duke of 111, 114, 179
Edwin, R. 81
Eldon Square, Newcastle
  redevelopment of 91–92
Elizabeth I
  proclamation of 1560 9
Elliott, D.M. 73
English Heritage 1, 73, 127, 128, 137–148, 153, 154, 156, 157, 158, 159, 160, 161, 162, 174, 178, 182, 183, 184, 185, 187, 190, 191
  advocacy role 144–145
  First Annual Report 142–143
  image 146
  membership 146

English Heritage *contd*
scope of 146–148
workload 143–144
Environment Committee of the House of Commons 1986 125–126, 149–155, 158
Esher, Lord 50, 58
and York 98–99
Essex Hall, Colchester 151
Europe
conservation in 25, 27–29, 55, 102
European Architectural Heritage Year 5, 55, 108, 110–115, 133, 187
funding 112
objectives of 111
organization 111–112
origins of 110–111
Euston, Earl of 73
Euston Arch 82
demolition of 84–85
Eversley, David 75–76

Faculty Jurisdiction Commission 124–125
Fawcett, Jane 3
Firwood Field, Bolton 112
Four Studies in Conservation
Bath 98–99
Chester 98–99
Chichester 98–99
York 98–99
France, *see* listing in

Galbraith, J.K. 66
Garton, W.J. 66, 69, 70
General Development Orders 39
Georgian Group 49–50, 51, 58, 78, 80, 81, 98, 187
Gilbert Scott, Sir Giles 14, 20, 146
Foreign Office and St Pancras Station 74
telephone kiosk 146, 147
Girouard, Mark 3, 15, 69, 179
GLC, *see* Greater London Council
Glendoran, Lord 75
Godrey, Walter 52, 65, 66
Gowers Committee 72–73, 149
Grade I listing 68, 134, 144, 157, 160, 184, 185, 205
definition 66, 199, 203
Grade II listing 68, 133, 134, 161, 184
definition 66, 199, 203
Grade II* listing 144, 157, 160, 184, 185, 205
definition 203
Grade III listing 68, 107
definition 66, 199, 203
Grade IV listing 69
Grafton, Duke of 139
Greater London Council 34, 76, 92, 108, 144, 160, 177
Greenwood, Anthony 94, 97, 98
Guildhall, Worcester 18

Hackpen Hill 23
Hadlow Court Castle, Kent 17
Hardwick, Philip 84
Harley Street, London 80
Hassell, J. 12
Harrington, Marquis of 41
Harvey, John 4, 20, 21, 27, 28, 89
Hawksmoor, Nicholas 123
churches of 47
HBMC, *see* Commission for Historic Buildings and Ancient Monuments
Heath, Edward 106
Hervey, Lord Francis 25
Heseltine, Michael 138, 143, 159
Hewison, Robert 181
Highclere, Hants 16
Highgate Cemetery, London 108, 109
Hill, Octavia 34
Historic Buildings and Ancient Monuments Act 1953 73
Historic Buildings Council 72, 73–76, 87, 90, 95, 96, 103, 133, 134, 135, 137, 139
Historic Monuments Act 1953 95
Historic Buildings and Conservation, see Circular 102/74
Historic Buildings and Conservation Areas, see Circular 8/87
Historic Towns: Preservation and Change 97, 98, 100, 101, 167
Hitchcock, Alfred Russell 82, 84
Hob Hurst's House and Hut 25
Hobhouse, Hermione 49, 108
Holford, Sir William 65, 66, 67, 73, 87, 89, 91, 96, 134, 201
Holland House, Kensington, London 57
Housing Act 1969 103
Housing and Town Planning Act 1919 37
Housing Etc. Act 1923 37–38, 55
Housing, Town Planning Etc. Act 1909 36–37
Howard, Ebenezer 36
HRH The Prince of Wales 114, 172
Hunt, Holman 20
Hunt, L. 12
Hunter, Michael 3
Hussey, Christopher 73

ICOMOS, *see* International Council of Monuments and Sites
International Council of Monuments and Sites 110–111
Insall, Donald W.
and Chester 98–99
Institute for Advanced Architectural Studies, York 3
Instructions to Investigators 66–69, 78, 107, 194–200
International Council for the Conservation of Monuments and Sites 3
Italy, conservation in 95, 96

Jenkins, Dame Jennifer 150
Jessel, Toby 139
Joiners' Hall, Salisbury 34
Joint Committee of the National Amenity Societies 144
Joint Urban Planning Group 87
Jokilehto, Jukha Ilmari 3
Jones, Brendon 66
Jones, Inigo
demolition of houses of 108
Jones, J. D. 90
Joseph, Sir Keith 86, 88

Keeble, Lewis 64
Kennet, Wayland 3, 89, 93, 95–96, 98, 100, 101, 102, 103
Kent, William 50
Kerr, Robert 21, 49, 75
Kipling, Rudyard 160

Lambarde, W. 9
Lancaster, Osbert 3
Lansdowne House, London 46
Lascalles, Sir Alan 73, 75
Lavenham, Suffolk 80
LCC, *see* London County Council
Legard, Charles 25
Leland, J. 9
listing and listed buildings and monuments, *see also* Advisory Committee on Listing 1, 3, 4, 12, 25, 28–29, 31, 32, 60, 65–70, 77, 78–79, 107, 119, 133, 158, 161, 178, 184–186, 194–200
and the 1944 Act 56–59
beginning of 65
classification 66, 78–79
criteria 167–168, 201–205
in France 12, 28–29, 95
interim lists 69–70
lack of advice to local government 77
Local Government Board 36–37
London County Council 34, 39, 47, 80, 81, 85, 177
Longton Town Hall, Stoke-on-Trent 151
Louis X, Grand Duke of Hesse 27–28
conservation decree of 1818 28
Lubbock, Sir John 23–25, 29, 36, 55
Lyons, D. 12

MacColl, James 96
MacCormac, Richard 179
Maclagan, Sir Eric 65
Maclagan Committee 67, 68, 167, 201
Macmillan, Harold 84, 85
Magnus, Sir Philip 65
Malrauz, André 95
Malton, T. 12
Managing England's Heritage: Setting Our Priorities in the 1990s 159
Mann, Sir James 73
Mansion House Square *see* No. 1 Poultry, London
Mappin and Webb site *see* No. 1 Poultry, London
Marlborough, Duchess of 10
May, Walter Barton 17
Mendelssohn, Eric 185
Merimée, Prosper 12
MHLG *see* Ministry of Housing and Local Government
Ministry of Health 37, 38, 55, 165
Ministry of Housing and Local Government 70, 73, 75, 76, 78, 87, 93, 95, 165
Ministry of Local Government and Planning 78
Ministry of Public Buildings and Works 73, 95, 104, 157
Ministry of Town and Country Planning 62, 67, 72, 78
Ministry of Works 72
Montpelier Row, Lewisham 81
Montagu, Lord, of Beaulieu 75, 142, 148, 157
Morrell, Philip Edward 36, 38, 55
Morris, William 19, 20, 36
and the SPAB 20–21
Morrison, W. S. 56
Mottistone, Lord 84
Murphy, Christopher 139

Nairn, Ian 85, 86
Nash, John 44, 48, 49
National Audit Office 157
Report 157–158
National Buildings Record 52
National Heritage Act 1983 138–140
National Heritage Committee of the House of Commons 161–164
National Heritge Memorial Fund 156, 162
National Lottery 163, 164
National Monuments Preservation Bill, 1873 23–25
National Monuments Record 52
National Trust 33, 34, 71–72, 74, 142, 182, 187
National Trust for Scotland 113, 114
National Trust Act 1937 71
National Trust Act 1939 71
Neale, J. M. 14
New College, Oxford 114
New Scotland Yard, London 108, 109
NHMF *see* National Heritage Memorial Fund
No. 1 Poultry, London 171, 172–176
Norfolk House, London 46
Northumberland House, London
demolition of 46–47
Norwich, Bishop of 121

Old Harlow, Essex 114
Old Post Office, Tintagel, Cornwall 34

Organisation of Ancient Monuments and Historic Buildings in England 136
The Way Forward 137–138

Page, Jennifer 157
Palmerston, Lord 49
Palumbo, Lord 172
Paternoster Square, London 89, 145
Patmore, Coventry 20
Pattison, Mark 20
Pearson, J. L. 14
Pembroke House, London
demolition of 4
Pennethorne, Sir James 47
Percy, Lord 25
Pevsner, Nikolaus 3, 21, 69, 70, 84, 102, 113, 134
Phillips, Hayden 157, 158
Piccadilly Hotel, London 44
Planning Acts *see also* individual Acts 4, 5
Planning Advisory Group 101
Planning and Compensation Act 187
Planning Bulletins 87, 88, 165, 167
Planning Bulletin No. 1 87, 97
Planning Bulletin No. 4 88
Planning (Listed Buildings and Conservation Areas) Act 1990 105, 187
Planning Policy Guidance Note PPG 1 169, 170, 171
Planning Policy Guidance Note PPG 15 128, 129, 163, 165, 168–171, 175, 178, 201, 204–205
Planning Policy Guidance Note PPG 16 2
Poole, Dorset 113
Preservation Policy Group 102–103
Protecting and Managing England's Heritage Property 157
Protecting Our Heritage 5
Pugin, A. C. 12, 14
Public Accounts Committe 157, 158, 159, 161

Redundant Churches Fund 123, 124, 127, 129
Regent Street, London 44
Rendel, Goodhart 66, 82
Rennie, John 47
Responsibilities for Conservation and Casework 156
restoration of buildings 4, 13
Victorian 13–15
Reynolds, Sir Joshua
demolition of house of 49
RIBA *see* Royal Institute of British Architects
Richards, J.M. 49
Ridley, Nicholas 96, 148
Robshaw, Peter 150
Rose Theatre, Southwark, London 145
Ross, Michael 2, 9
Rothschild, Lord 163
Rowlandson, T. 12
Royal Academy 47
Royal Commission on Ancient and Historical Monuments, 1908 29, 30, 34
First Report 30
Royal Commission on Historical Manuscripts, 1869 29
Royal Commission on Historical Monuments 67, 69, 136, 137, 156
Royal Fine Art Commission 34, 78, 121, 134, 156, 163
Royal Institute of British Architects 21, 47, 94
Royal Opera House, Covent Garden 145
Royal Society of Arts 34
Royal Town Planning Institute, 133
Ruskin, John 14, 16
and anti-restoration 16–21

St Bernard's Hospital, Ealing 151
St Katherine's Dock, London 114
Salisbury, Marquess of 55, 60
Salvin, A. 14, 15
Samlesbury Hall, Lancashire 114
Sandford, Lord 105
Sandys, Duncan 88, 89–90, 91, 93, 96, 111
SAVE Britain's Architectural Heritage 115, 125, 174, 177, 187
scheduled monuments *see also* listed buildings 1
secteurs sauvegardés 95
Settle, North Yorkshire 112
17 Fleet Street, London 34
Shambles, The, Manchester 106
Sharp, Dame Evelyn 89, 90, 94
Sharp, Thomas 77
Shaw, Giles 139
Shaw, Norman 44
Silbury Hill 23
Silkin, Lord 59
Sisson, Marshall 67
site classé 95
Sitwell, Osbert 48
Skelmersdale, Lord 126
statement of 126
Sladen, Tesesa 150
Smith, J.F. 3
Smith, W.H. 25
Soane, Sir John 44
Society of Antiquaries 9, 12, 13, 17, 18, 23, 27, 55
Society for Photographing Relics of Old London 32
Society for the Preservation of Ancient Buildings 3, 14, 16–21, 23, 24, 36, 43, 47, 49, 50, 51, 55, 58, 75, 139, 187
SPAB *see* Society for the Preservation of Ancient Buildings
Spitalfields Market, London 145
Sproule, Anna 49
Stanhope, Earl 18

Stevens, Jocelyn 148, 156, 160, 161, 162
Stirling, James 172
Stonehenge 25, 26, 27
Stowe, John 9
Stratford Place, London 80, 81
Strong, Roy 113
Suddards, Roger 4
Summerson, Sir John 52, 62, 63, 64, 65, 69, 73, 87, 133, 134
Survey of London 34, 39
Sykes, Christopher 46

Tadema, Alma 20
Talman, William 183
Taylor, George Ladwell 17
telephone kiosks
  colour of 78
  listing of 146–147
Thames Street, Dorchester 114
Thorpe, T. 14
Tower of London 145
Town and Country Amenities Act 1974 107
Town and Country Planning Act 1932 38–41, 56, 57, 119
Town and Country Planning Act 1944 55, 56–59, 67, 119, 185, 189
Town and Country Planning Act 1947 2, 31, 43, 55, 59–61, 64, 65, 67, 119, 169, 185, 189
Town and Country Planning Act 1968 101–102
Town and Country Planning Act 1971 104, 105, 106
Town and Country Planning Act 1990 187
Town and Country Planning (Amendment) Act 1972 104–105
town grants 74
Town Planning Institute *see* also Royal Town Planning Institute 47
tree preservation orders 39
Twentieth Century Society 187

UNESCO 110, 111
Unwin, Sir Raymond 55
Uppark, West Sussex 182, 183
  restoration of 182

Van der Rohe, Mies 172
Vanbrugh, Sir John 10–12, 15
Vane, W.M.F. 73
Victoria County Histories 34
Victorian architecture
  demolition of 82–85
Victorian Society 3, 83, 150, 151, 187

Walker, Peter 134
Warwick, Earl of 58
Waterloo Bridge, London
  demoliton of 47
Webb, Benjamin 14
West Kennet 23
West Wycombe, Berkshire 34, 112
Westminster Abbey 13, 122, 123
What is Our Heritage? 114
Wheeler, Sir Mortimer 121
Wilding, Richard 127–128
Wilding Report 127–128
Williams-Ellis, Clough 44
Wilson, Harold 92, 105
Windsor Castle 5
  fire 178–181
Winslow Hall, Buckinghamshire 79
Woodstock Manor 10
Wood, Alfred A. 112
Wood, Anthony A. 9
Workskett, Roy 97–98, 177
Worsley, Giles 179
Wren, Sir Christopher 5
  churches of 20, 47
Wyatt, James 13–14
Wyatville, Jeffrey 179, 181

Young, Sir George 169

zones protégés 95

Printed in the United Kingdom
by Lightning Source UK Ltd.
124334UK00001B/67/A